THE Nth POWER

EXPLAINING THE UNEXPLANABLE
"WHAT ARE GHOSTS, REALLY?"

Dr. Mitchell Wick

authorHOUSE®

AuthorHouse™
1663 Liberty Drive
Bloomington, IN 47403
www.authorhouse.com
Phone: 1 (800) 839-8640

© 2018 Dr. Mitchell Wick. All rights reserved.

No part of this book may be reproduced, stored in a retrieval system, or transmitted by any means without the written permission of the author.

Published by AuthorHouse 09/05/2018

ISBN: 978-1-5462-4820-0 (sc)
ISBN: 978-1-5462-4819-4 (e)

Print information available on the last page.

Any people depicted in stock imagery provided by Getty Images are models, and such images are being used for illustrative purposes only.
Certain stock imagery © Getty Images.

This book is printed on acid-free paper.

Because of the dynamic nature of the Internet, any web addresses or links contained in this book may have changed since publication and may no longer be valid. The views expressed in this work are solely those of the author and do not necessarily reflect the views of the publisher, and the publisher hereby disclaims any responsibility for them.

FOREWORD AND INTRODUCTION

THIS AUTHOR IS A SCIENTIST. I'm not from Missouri but like most Missouri-ites I have to be shown. There is ample anecdotal evidence which is presumptive evidence of the existence of ghosts or other para-normal phenomena. If one uses the Scientific Method to verify the hypothesis that Ghosts are real and not just visual or auditory hallucinations one needs to have hard empirical data documenting phenomena which otherwise can't be explained by most types of hard science. I am a Physicist and a Physician with an extensive mathematical and scientific background and I will attempt to prove that paranormal phenomena exist which cannot easily be explained by physics.

CONTENTS

Foreword And Introduction.. v

CHAPTER 1
Spooky Action at a Distance Electrons or Photons Pair from
Great Distances while Exchanging Information................................. 1

CHAPTER 2
Evidence that Schrodinger's Paradox is that the Presence of an
Intelligent Observer can Change Reality 7

CHAPTER 3
Mathematical Proof of 'Magic' or it's Equivalent............................. 10

CHAPTER 4
A Mathematical Solution to the Mass Gap Problem in Yang
Mills Theory 13

CHAPTER 5
A Short Essay to Illustrate that a One -Inch Equation Is "The
Equation of Everything" 16

CHAPTER 6
The Wave Properties of Matter and Spacetime By Dr. Mitchell
Albert Wick... 22

CHAPTER 7
How Apparitions May be Echos from Other Space-Time

Manifolds Reflecting as Mirror or Echo Dimensions
Overlapping with our Spacetime Manifold and Dimensions............. 27

CHAPTER 8
How Can Time Travel Backwards Occur Without
Incorporating a Force Equivalent to the Big Bang or Big Crunch?... 29

CHAPTER 9
How Heuristic Algorhythms Used for Artifical Intelligence
Can Complete the Ontologic Proof(Mathematical Proof of
God as the Comilation of the Masssive Higgs Field and the
Tachyon Where the Tachyon is the Brain and the Higgs Field
is the Body.).. 35

CHAPTER 10
How Did Life Initially Begin? 49

CHAPTER 11
The Final Unification Equation of Quantum Mechanics and
Relativity 56

CHAPTER 12
The Relationships of Black Holes and the Number of Extant
Universes... 58

CHAPTER 13
What are the Chances of a Paradox in Space-Time in the Future?.... 62

CHAPTER 14
What is Hyperspace?.. 64

CHAPTER 15
The Big Bounce 67

CHAPTER 16
What Came First; The Big Bang or the Formation of the
Multi-Verse? 75

CHAPTER 16.1
How Did Gravity Form? 78

CHAPTER 17
The Big Bang 80

CHAPTER 18
What is the Configuration of the Multi-Verse? 83

CHAPTER 19
A Short Essay Regarding the Equation of Everything 87

CHAPTER 20
Physical Cosmology; Isotropism VS. an Isotropism of our Universe 92

CHAPTER 21
How the Equation of Everything Applies to the Big Bang 100

CHAPTER 22
Information Exchange Via Spooky Action at a Distance 104

Appendix 107
Glossary 111

CHAPTER 1

SPOOKY ACTION AT A DISTANCE ELECTRONS OR PHOTONS PAIR FROM GREAT DISTANCES WHILE EXCHANGING INFORMATION.

Electron pairs were scientific show to exchange information such as spin across great distances whereby even at the speed of light the information exchange wouldn't occur as fast as it does. Also this information exchange is altered by observers doing measurements. According to the C.P.T. (Charge, Parity, Time)Theorem the information exchange can reverse parity or spin and possible time while retaining the negative charge over parsecs. In terms of the information exchange space-time may have to be warped or perturbed in such a way as the two electrons appear to be at great distances and at the same time be spacially adjacent. This can occur if space-time is folded or hyperfolded by the information exchange such as spin or by the presence of the intelligent observer measuring the two electrons. Electrons travel at approximately 2/3"c" and have a minuscule mass causing a miniscule change in the effect of gravity or spacetime curvature so if space-time curvature from two isolated electrons is very slight, how can it be curved to such a degree as to cause a folding between the electrons. Electron effects travel out concentrically as an electron gas throughout space even incorporating with the Background Microwave Radiation of deep space with the energy density of a near vacuum as shown by the Cosmologic Constant. As electron gasses travel at near 3×10^8 meters/sec(2/3) the inertial mass of the electron would begin to increase due to special relativity and this in turn would increase space-time

Dr. Mitchell Wick

curvature with dilating time beginning to constrict space not to zero but definitely constricted. The spin moment of each electron would spin the space as it constricts and CAN CAUSE A FOLDING OR WARPING OF SPACE IN DILATING TIME DUE TO THE VELOCITY OF EACH ELECTRON BEING AT 2/3 the speed of light. Ghosts may be manifested as low energy electromagnetic radiation which is extremely low frequency long amplitude waves; but as electromagnetic radiation the mass would be that of photons 3x10-18eV/c^2. They would also travel at velocities approaching "c" which is in constricted space and infinitely dilated time which would increase the inertial mass of the photons into a continuous energy stream. This too would also have the effect of warping space and as light and other electromagnetic radiation travel in almost a straight line (they are bent by gravity and any heavy masses like black hole). A LASER EFFECT CAN BE APPLIED TO LOW ENERGY PHOTONS AS WELL. Also note that there is a drop in potential difference or voltage (kVp) in the region of ghost sightings. Voltage measures the velocity of electron streams between positive and negative terminal whereby electrons travel from the negative to the positive and are affected by magnetic moments. A drop in the potential difference means that the electron stream from negative to positive poles is reduced which makes the velocity of the electrons reduce while the intensity (I) or the number of electrons remains the same. Also a drop in temperature measures a loss of heat in kilocalories of the measured system. Using the equation of everything $!(n)abcd = \div\ n = 0$ to ∞ eigenstates of $2^n + 1\pi\omega$ i to

$$\frac{j}{2^n} + 1\pi\omega\ j \to i + 2\pi \left[R\ abc + or - \frac{1}{2R} g \frac{ab}{+} or - ih\Lambda\ ba\ \rho\ ab \right]$$

the mass of electromagnetic radation is $3 \times 10 - \frac{18kV}{c^2 w}$ hichmakes a very large number for the right side of the equation or + or – i (6.63×10^{-34} joule – sec(10^{-110}) (ρ ab) where $\rho\hbar$ = R ab and R ab = 10^{54} kg mass of the universe so its energy equivalent isis 10^54(6.64×10^{-34} = 6.63×10^{20} joules. So consequently $\frac{1}{i(6.63\times 10^{-34})(10^{-110})(6.63\times 10^{20})} = i(37\times 10^{124})$ as $\frac{1}{i} = i$. There is negligible gravity as the mass is miniscule. The spiral operator zeros out and the left side has an infinite number of eigenstates and as ghosts would smear across curved space0time every eigen-state except zero could have ghosts so we get i(37×10^124(='~infinity) this result is incomplete and does not rule out the existence of ghosts but doesn't prove it either. A nonsense result which would have ruled out ghosts would have been 0=infinity, but this wasn't the case. Ghosts are vorticies whose center is the point of intersection of two space-time manifolds moving in near equal but opposite directions on Cartesian Coordinates. They are acting as a carrier wave to electromagnetic radiation. This manifests as two dimensional or holographic projections as from a projector which is the vortex emerging from the central point or nexus of the two manifolds where the nexus point is through the 11th dimension. As a super-symmetrical throat and it requires a tuning up of the temporal lobe of the brain and the micro-tubules which act as receptor. The path is determined by the action formula S-S=-1/2k^2[(d^11(-g)]^1/2R2-R1 where this R2-R1 are the regions of the two space-time manifolds curvature variants caused by the inertial mass of the curving matter oof each manifold minus the minus the miniscule mass perturbation caused by the holographic projection from the vortex of the space between the manifolds or universes along the 11th dimension. Also, nexus junctions with a phase shift between of 11 to 15 degrees allow spectres from the shadow dimensional planes in the shadow universe and accounting for the missing mass in our universe who se inertial mass should slow down the accelerated expansion of our galaxies but doesn't because the missing mass isn't in our perceivable dimensions. Also nothing doesn't exist never existed and never will exist. Ghosts are not nothing therefore ghosts must exist Even if a construct of the frontal lobe of each intelligent brain is required to perceive ghosts though the microtubules coating the frontal lobe as determined by the University of Arizona Neurology department in determining "life after death" experiments they still haven't ruled out the true existence of paranormal phenomena as opposed to visual or auditory hallucinations. LOGIC PROOF OF PARANORMAL PHENOMENA. Everything is in the set of all elements and properties of nature or A∞ *and the null set is nothing {}. All the elements of nature* is {a1,a2,a3...an→a∞} *where the union of a n includes that phenomena of paranormal phenomena. As paranormal* Phenomena have properties which are observable such as a

Dr. Mitchell Wick

drop in KV p and temperature these phenomena which are all inclusive of any and all observables linked with the phenomena make them a subset of the infinite set of all phenomena in nature both observable and not observable such that *{a n ϵ A∞} and a n is not an element or property of {} or the null set; which has no properties.* Therefore from a logic standpoint paranormal phenomena do exist at least in certain definable quantum states and the spiral operator of the equation of everything is $\prod n = 0$ *to* ∞ *eigenstates of* $2^n + \dfrac{1\pi}{2^n \pi}$ *is appraching* $\dfrac{\circ}{\circ}$ *which is everything except zero over an near infinite number of eigenstates.* As long as paranormal phenomena occupy one or multiple eigenstates the spiral operator operating on angular momentum includes eigenstate with paranormal phenomena.

Note utilizing deBroglie Wavelengths there is a corresponding wavelength and frequency for each and every mass not matter how 'large or small. The equation $\lambda = \hbar \dfrac{}{mc}$ or $h \dfrac{}{2\pi(mv)^{-1}w}$ *here v < c is nonelectromagnetic radiaition is used. Of course as photons comprise electromagnetic rad???* Has a miniscule mass of 3×10^-18 ev/c^2 that mass is propelled with a velocity of everything from 3×10^8m/sec to ~0 as an asymptotic function as photons cannot be stopped as absolute zero is approached. Momentum is mass times velocity and the energy of a photon E=hv *where v is the corresponding frequency to the energy level of that photon.* Everything with Mass has momentum including energy so the vacuum state is a false vacuum state as it has potential energy from leptons and gluons making up the fabric of space with a miniscule amount of kinetic energy to separate the planes that are totally parallel before they insetsect to form dimensions. Paranormal phenomena have a corresponding frequency and wavelength whether they be K-suryon Waves (see What is the Dimension of Time by this author) or other types of low energy electromagnetic radiation;and as photons they can travel at all speeds except zero and Spooky Action at a Distance can explain the phenomenon of "bilocation" whereby an individual could physically be in one place and interact with people or objects in another place. "Bilocation is considered a paranormal phenomenon but so could "Spooky Action at a Distance" and when one considers that everything has a frequency and wavelength the wavelength of the observer MUST INTERACT WITH THE WAVELENGTH OF THE PARANORMAL PHENOMENON which changes the information flow that is perceived by the observer and translated in visible and audible impressions by the observer. If consciousness can be determined to be energy which anecdotally is very true(by out of body experiences)then conscious will leave the defunct or even comatose bodies via the microtubules aligning the cerebral cortex of the frontal lobe as determined by the Neurology Dept. at the University of Arizona. Can consciousness exist without life? One would think the answer would be no as heavy anesthesia reversibly destroys consciousness and a sense of time in a dreamless state;however the brain can reproduce living experiences from the past with simple pressure to the correct gyri and sulci of the brain as in the Pennfield Experiments of 1958 which were used to treat epilepsy. Still K-suryon waves do exist with a mass,frequency and wavelength and are considered consciousness waves of a sort. So if K-suryon waves have a specific energy level and are consciousness waves then consciousness is energy and ghosts or paranormal phenomena are possible. What happens to these K-suryon waves when a patient is in a deep coma or anesthesia? The wavelength increases dramatically corresponding to alpha waves in an electroencephalogram;this forces a high frequency or high energy level for K-suryon waves which can cause a divergence of consciousness away from the corporeal body and into space. HOW DO A DROP IN KVP AND TEMPERATURE DROP RELATE TO PARANORMAL PHENOMENA AND DIMENSIONAL SLIPPAGE. If there is a negative mass involved with two dimensional or holographic like images of low energy electromagnetic radiation in the radiowave band;then there would be a contribution to the expansion of space-time from the negative mass as well as a local anti-gravitational effect which is larger than the Cosmologic Constant. An increase in the space which is expanding geometrically would lower the ambient temperature slightly as a local cold spot and the KV p would indeed drop and the resistance of the space(which has a miniscule mass)would slightly increase. Despite this negative mass has never been recorded and it is unknown or unclear if there's a decrease in gravity around the region of "ghost sitings". The phenomenon of "Poultergeists" display anti-gravity phenomena with objects flying about with no reasonable explanation although anti-gravity spots would explain them.

CHAPTER 2

EVIDENCE THAT SCHRODINGER'S PARADOX IS THAT THE PRESENCE OF AN INTELLIGENT OBSERVER CAN CHANGE REALITY

Every science student that has taken either Quantum Mechanics and Calculus Physics is probably acquainted with the Schrodinger Equation and Schrodinger's Cat. In essence the cat is in a box and is experiencing what is known in paranormal terms as "bilocation"in which case an object or individual which is alive or was alive can be in two locations at once or at least it's consciousness. How can a comatose individual also be present thousands of miles away doing activities with other people and conscious? Remembering the definition of impossible is" that which is beyond the comprehension of man"and bi-location seems impossible unless the bilocated individual was identical twins. If Schrodinger's Cat was alive and dead at the same time depending on whether or not the cat was in the box and whether or not there was an intelligent observer,then Schrodinger's Cat would be experiencing bi-location which has been reported over the years but never had any documented evidence as to why it occurred. How does a measurer or an observer whether intelligent or not change reality? In the case of Photon Pairing relating to Spooky Action at a Distance or electron pairing; the spin or an electron can have a measurable change as to whether it is being observed and measured or not. Does having an intelligent observer change the reality of Schrodinger's Cat? Many people who have been highly trained and are affiliated with some major institutions state that the presence of an intelligent observer can change reality. There is actually a theory purported by a physicist and acknowledged by others including Neil deGrasse Tyson(head of the Hayden Planetarium in NewYork)that our universe is analogous to a computer program(similar to a movie plot in the late 1980's). This would only be true if our universe is considered analogous to the two dimensional world sheet as purported in string theory. A two dimensional universe would have length and width but depth would only be Planck Length or 10^{-33}cm although a computer screen with voxels and pixels are flatter. The intelligent observer could be anything anywhere with deterministic intelligence but will also need the expertise to download a universe (one in between 2 billion and 10^{100} or a googolplex. As the Higgs Field is an integral part of this any other universes as part of the First Event; it cannot be separate and as not separate it is interacting with all the extant universes including ours. This plus Stephen Hawking premise that intelligent life is at least three and probably n dimensional rather than two in order to be alive makes this idea unlikely; although in Quantum Mechanics every possibility exists so there may very well be a universe that fits those criteria although not our universe. Sound waves are formed by vibrations in a medium at different frequencies that follow the Doppler Effect and the need for a receiver is NOT A NECESSARY CONDITION FOR SOUND WAVES TO EXIST. Therefore if a tree falls in a forest and nobody is around to hear it still makes a sound as long as there is a medium for the sound waves to be transmitted and since a receiver isn't necessary for the sound waves to be produced an intelligent observer doesn't change that reality.

Supernovae do not need observers to exist and neither do stars. nebulae or galaxies. Indeed there is a multi-verse which hasn't been observed or measured yet although proven mathematically;yet they exist even if comprised primarily of black holes due to their extreme age. So why are measurements changed in electrons by the presence of a measurer?The measuring device that is part of what's being measured is skewed. Since the electron clouds of each and any electron can have clouds existing over parsecs or greater;and the information between two electrons can be exchanged through these clouds even over parsecs. The total spin of the electron cloud must be measured outside of the influence of that electron cloud;as of this date such technology does not exist so the measurements are skewed as the size of the electron cloud communicating information such as spin with the other electron cloud of the pair cannot be exactly measured;indeed Quantum Mechanics states that there are no exact measurments but if the whole system of both overlapping electron clouds are considered over parsecs of space;then the observer is in fact not changing reality but is lacking necessary information or measurements to get an accurate answer.

CHAPTER 3

MATHEMATICAL PROOF OF 'MAGIC' OR IT'S EQUIVALENT

When one uses the Ontologic Proof that the Higgs Field with Tachyons as the Permanent semi-radius tachyons(Kaku 2016)with the Tensor Virial Theorem to prove the existence of that which is "one with everything"as a massive vortex following the space-time continuum which is also a vortex or whirlpool effect of the perfect fluid continuum(without beginning or ending)one does not conclusively mathematically prove thought although the tachyon can and does act as though it is the brain and spinal cord of the Higgs Bosonic Field which is the body. In the book "What is the Dimension of Time?Time,Mass and Energy the Hows and Whys?"it is shown that the rogue tachyon which triggered the Time Oscillation Paradox which had a 100%probability of being correct in infinitely dilated time is the POINT OF CREATION. Whether or not this rogue tachyon which dropped to "c"from greater than "c"velocity is a deterministic act or not is thus far unclear and unproven although K-suryon waves have been determined to be consciousness waves with a unique frequency and energy level which can match up with paranormal phenomena or ghost siteings picked up by the microtubules of the frontal parietal temporal and occipital lobes of the brain where in this case the microtubules act as receivers of K-suryon waves which are transferred in the microtubules of the occipital lobe into a visual orb or vortex or may trigger memory impulses which match with past past experience to cause "materialization". Is consciousness energy?According to K-suryon waves the answer is yes;yet people under deep anesthesia or comas have to illustrate a traveling of consciousness from the body elsewhere. Unfortunately,this is an explanation of the Paranormal Phenomenon known as Bi-location in which one's body is in one place and the image or consciousness is in another place. This bi-location phenomenon an explain ghosts if the donor body is still alive and can even explain "astral projection"such as when patients having C.P.R. can view their own bodies from another location and report it after returning to consciousness...alive. So how can "withchcraft"or magic be proven. The Equation of Everything has 524,288 permutuations and are all inclusive. When the energy density of matter for mass and space-time as the dual vector field of Poisson'sEquation equals the wave function of a point particle at time t all the different combinations of energy can be constricted by constricting space-time such as at the event horizon of any active black hole. As unusual as it may sound since the time independent case of the Schrodinger Equation has the wave function of point r equals the energy density of matter plus the cosmologic constant(which approaches zero as the event horizon is a black hole is approached). If astral projection or bi-location using K-suryon waves can travel via Spooky Action at a Distance toward any active black hole or combination of black holes the energy density of matter can be translated into the wave function of any point which has no scattering due to constricted space-time. In a way this is analogus to a "generator of magic". Of course in trying to prove this phenomenon in any way that isn't mathematical is past our present technology and may continue to be so until the Omega Point is reached(when everything that is learnable is learned).

An Detailed Explanation of the Equation of Everything by Dr. Mitchell Albert Wick

The Riemannian or Lorenzian Curved Space-time which is the Circumference of the compactified circle representing type IIA string theory and Heterotic 8×8 string theory is the Region of curved space-time as a tensor of the fourth degree over n eigen-states of energy from n=1 to infinity-epsilon which is a small amount. n=0 eigenstate of energy doesn't exist as energy cannot be created or destroyed whether it be potential energy, kinetic energy, heat the Strong Force, weak force representing nuclear decay or electromagnetism. THIS FIGURE EQUALS THE INFINITE PRODUCT FROM n=1 to infinity-epsilon eigen-states of energy of the SPIRAL OPERATOR OR 2^n+1(pi)/2^n(pi) operating on the function of angular momentum (which also has a Law of Conservation stating that angular momentum cannot be created or destroyed). The numerator goes from i the initial event to j the final event or the spiral increasing from a point with infinite curvature to a flat infinite diameter circle as the final event in the numerator and the denominator is from j to i or the final event to the initial event with the spiral going from an infinite diameter circledown to an infinite curvature point. This results in the expression infinity/infinity=everything except zero proving that nothing doesn't exist and result in a near infinite number of constants which are added to R abc which is Euclidian or flat space being acted upon or curved either inward (gravity) or outward (antigravity) for the metric (anything measured) g ab so the topological region R resulting from the metric g ab results in R g ab with an inertial mass of R ab or the region for the Ricci Tensor of inertial mass for the metric g ab. This metric goes into the denominator so R abc+ or -1/2R g ab/R ab. Time of course is gravity or anti-gravity/mass. R ab=$\hbar\rho$ ab and the heteotic property of flipping between dimensions due to Wick Rotation results In i$\hbar$ ($\rho + \Lambda$) where ρ is the energy density of matter from Poisson's Equation. $\hbar = h\dfrac{}{2\pi}$ and the 2π goes into the numerator making Circumference = 2π (Radius) where the + radius is + Riemann forces and − radius is − Riemann Forces which is R abc + or $-\dfrac{1}{2}g\dfrac{ab}{i}h(\rho\ ab + \Lambda\ ba)$ and the circumference is curved Riemanniam or Lorenzian Space − time and the compactification is a circle or string *theory with an infinite number of CONSTANT???* CONSTANTS ADDED TO THE 2(pi)radius with ANGULAR MOMENTUM FUNCTION OF THE SPIRAL OPERATOR IS THE COMPACTIFICATION PROCESS OR CURLING UP THE FUNCTION IN THE DENOMINATOR RESULTING IN A CIRCLE. WE ALREADY KNOW THAT SCHRODINGER'S EQUATION HAS ih(bar) partial derivative of the wave function of a point particle=ih(bar) with respect to time in the time dependant case which equals ih(bar) the derivative of the tensor sum or Christoffel symbol(rho+cosmoloigic constant)

CHAPTER 4

A MATHEMATICAL SOLUTION TO THE MASS GAP PROBLEM IN YANG MILLS THEORY

To Prove that for any compact simple gauge group G, a non-trivial quantum Yang-Mills theory exists in $\mathbb{R}^4$ *and has a mass gap* $\Delta > 0$. The Yang-Mills (non-Abelian) quantum field theory underlying the standard model of particle physics in $\mathbb{R}^4$ *euclidian space (flat space) has a mass gap* Δ *which is the mass of the least massive particle predicted in the theory... The Mass gap is the* difference in energy between the vacuum and the next lowest energy state. The energy of a vacuum is 0 by definition and assuming that all energy states are particles in plane-waves with the mass gap being the mass of the lightest particle. For any given field $\phi(x)$ *the theory has a mass gap if the two point function has the property of* $<\phi(0, t)\phi(0,0) = \Sigma n\, A\, n\, e^\wedge - \Delta nt)$ with $\Delta o > 0$ being the eigenvalue for the lowest energy in the spectrum of the Hamiltonian and the mass gap. In lattice this mass gap occurs. Photons and the Mass gap in a Lattice Regardless of the U(1) Yang Mills non-Abelian gauge symmetry group which incorporates electromagnetic radiation; the photons must be extended to close the mass gap as in the Omega Particle. At extremely low temperatures and energies photons can change their state as in matter as they do have a non-resting mass. In black holes there are 252 postulated states of matter according to Stephen Hawking causing electromagnetic radiation to express other states of energy/matter such as Boso - Einsteinian Condensate As a condensed form of radiation photons can get rapped in a lattice pattern. In terms of gauge symmetry with

Dr. Mitchell Wick

no resting frame the mass of a photon is <10^{-18} electron volts/c2. The spin is 1, parity is -1, charge(q)<10^{-35} e and photons are stable during an indefinite period of time. The mass of a photon is such that the energy squared=momentum squared (c^2)+mass squared c^4 in other words $E^2=p^2c^2+m^2c^4$ where m→0. *Photons follow the CPT Theorem of charge, parity and time where reversal of time doesn't affect the outcome...* Also the direction component of motion is $\pm\hbar$ *which is helical.* As there is no rest frame there is no resting mass for photons although they are attracted or bent by gravity and therefore curve space-time as at the Event Horizon of a Black Hole. Condensed the gauge symmetry notation for intrinsic properties of photons would be $1(J^{pc})=0,1(1--)$ where p=parity which is -1 and c is charge which is also -1. Spin is 1 and mass is listed as 0 in terms of gauge symmetry... In addition the C parity which is charge parity is also -1. So with a gauge symmetry of $1(J^{pc})=0,1,(1--)$ and in the Mass gap equation the photon would be $<\Phi(0,t)\Phi(0,0)>$ *becomes* $<(0,1(1--)\Phi(0,0)>\rightarrow \Sigma \, Ane^{-\Delta nt}$. *In* corporating this in to the Ising Model with regard to Conformal Field Theory

G/H=S U (2)k × SU(2)1/SU(2)k+1 where C G/H =3k/k+2+1-3(k+1)/(k+1)+2=1-6/(k+2)(k+3). Minimal unitary models exist such that m=k+2=3,4,5 and the unitary series is SU (2)k × SU(2)1/SU(2) k+1. Ward like identities can be used for rational conformal field theories or correlation functions. The Knizhnik Zanrolodchikos Relation gives explicit expression of correlation functions where $k\frac{\partial}{\partial zi}-\Sigma j=\frac{\bar{i}ti^{atj^a}}{zi}-zj)<,g(z1,z'1)...g(zn,z'n)\geq 0$ where i doesn't equal j and i is the initial event and j is the final event.

In terms of Conformal Field Theories which measure mass in terms of conformal gravity we find the Super-conformal Minimal Series $|h+->=S+-(0)|0>$ where $|h->=Go|h+>$ where $|0>$ is the bosonic vacuum state S+-(Z) is the spin field the conformal field is $\phi h(z, 0)$ and $|h\geq \phi h(0,0)|0>$ where the highest weight vacuum $|h>$ is annihilated by the generators with positive indicies. The vacuum state is a Fermionic field where $|h+->$ is a fermion. $\Gamma = (-1)^F$ where F = Fermion number. Vacuum divided to $\Gamma|h\pm\geq\pm|h\pm>$ where $\{\Gamma, Gn\} = \{\Gamma, Ln\} = 0$. G 0 is the ground state of energy with regard to conformal gravity $G0^2 = L0 - \frac{c}{16}$ where $\frac{G}{H} = \frac{SU(2)kxSU(2)2}{SU(2)k} + 2$ in Yang Mills Format where H is a primary conformal field Hamiltonian for k to n eigenstates of energy. In the 2 dimensional lattice Ising Model Equation the conformal weight Δ of the field whose energy operator products $\epsilon n = \sigma n \sigma n + 1$ where σ = isospins from the energy field. $<\epsilon n \epsilon 0 \geq x^{-2\left(d-\frac{1}{\nu}\right)w}$ here ν and η are critical exponents for the Ising Model for the spin field and energy

Where h+h' are the Ising Fields for Conformal Field whose Conformal Gravity is Δ such that $gij = x^{-d} + 2 - \eta$ where d = dimension which approximately equals $x^{-2\Delta}$ where the critical exponents are $\eta = \frac{1}{4}$ and $\nu = 1$ so in the minimal model $= 3\sigma = \frac{\phi 1_{16,1}}{16}$ and $\epsilon = \frac{\phi 1_{2,1}}{2}$. In this case the mass $<\phi(0,t)\phi(0,0) \geq \Sigma nA\, n\, e^{(-\Delta nt)w}$ here $gij = x^{-2\Delta}$ where Δ = conformal weight or mass. As $gij = x^{2\Delta}e^{-x(\epsilon)w}$ here ϵ is the correlation length at cirat criticality goes to ∞ so $e^{-x}\epsilon$ becomes $\frac{1}{e^o}$ or 0. Zero (0) at critical temperature causes the system to lose all dependence on fundamental length and the mass gap or conformal gravity gap $-\Delta n$ holds. *This* is the lowest energy in the spectrum of the Hamiltonian and photons at 0,1(1--)OR 1(I^pc)ARE THE NEXT LEVEL RENDERING FERMIONIC FIELD WITH PHOTONS AT CRITICAL LENGTH AND KEIGENSTATE FOR GROUND STATE AND K+1... EIGENSTATE FOR ENERGY OF PHOTON $\epsilon 0\epsilon 1 = \Sigma K \rightarrow K + 1$ FOR $\epsilon 1$ PHOTON ENERGY STATE. D is expanded to four dimensional Euclidian space in $x^{-2(d-1/\nu)}$ where $d = \mathbb{R}^4$ rendering $gij = \frac{x^{-2(\mathbb{R}^4-1)}}{\nu}$ where $\mathbb{R}^4 - \frac{1}{\nu}$ is conformal weight Δ in Euclidian 4 space and therefore the mass gap Δ where $\Delta > 0$,

CHAPTER 5

A SHORT ESSAY TO ILLUSTRATE THAT A ONE -INCH EQUATION IS "THE EQUATION OF EVERYTHING"

Author: Dr. Mitchell Albert Wick

The short simple relationship in verbal terms space-time is directly proportional to space and inversely proportional to mass is in reality when converted to metric tensors an equation which compactifies to a circle as does type IIA string theory and possibly all five dual string theories comprising M Theory. This equation has over 500,000 permutations which may in all actuality be all the equations in Nature including those discovered and undiscovered.

THE TENSOR FORM OF SPACE-TIME=SPACE/MASS

Space-time constricts as in Schwarzchild Space-time as the event horizon of any active black hole is approached. Black holes are extremely dense with extremely compressed mass and the gravitational effect or perturbation is extreme causing the observer to note a "frozen" image in time in as much the event horizon appears to be dilated time (time slowing down near heavy masses) with the constrictive effect of progressively increasing space-time curvature as the event horizon is being approached although an infinite curvature point of space-time is never reached(asymptotic). Space-time is inversely proportional to mass.

The Nth Power

Based on the Line Element space-time or ds^2=dx^2+dy^2+dz^-c^2dt^2+dr^2 indicates that even with the Relativistic effect on the dimension of time noted by c^2dt^2 space-time is direct proportional to space.

In terms of Metric Tensors Curved Riemannian or Lorenzian Space-time as a tensor of the fourth degree in n-dimensional space where n is from 1 toward infinity without reaching it. Riemann has 256 permutations between covariant and contravariant tensors with the Bianchi Identity applied and has the properties of being anti-symmetrical and Abelian.

This family of values is determined to be equal to a SPIRAL OPERATOR with regard to the function of Momentum producing a family of constants which are added to the equation of Flat Euclidian Space plus or minus the space-time curvature metric known as the effects of gravity and anti-gravity(which incorporates the Cosmologic Constant or the effects of Dark Energy. This value is the vector or tensor product of the reciprocal of inertial mass which is determined in this case by the Ricci Tensor. The equation is $\mathbb{R}(n) = \prod$ (from n = 0 to a eigenstates of energy) of the spiral operator $(2^n+1)\pi P \; i \to \dfrac{j}{2^{n\pi P_j}} \to i$ where i = the initial event and j the final event indicating a bi – conal configuration which is asymptotic to a spiral operating on momentum(P) as heavy mass Such as black holes in the denominator and toward the v=near vacuum of deep space and indicated in the numerator. These will result in a family of constants which are finite as eigen-state infinity is excluded as it results in the spiral operator zeroing out as an infinite curvature of a point as would only occur in an imploded universe such as a "Big Crunch".

This nearly infinite number of constants is added to R abc(flat Euclidian Space)+ or –1/2R g ab(space-time curvature metric or effect of gravity curving space-time inward and anti-gravity flattening space-time out such as the Cosmologic Constant being acted upon by the metric g(any measured quantity according to the Action Formula.) This is the dot product with regard to the reciprocal of the Ricci Tensor R ab which

Dr. Mitchell Wick

would normally be multiplied by a constant such as ~ $1/c^2$ to exactly form the Energy Density of Matter ρ ab. Due to the deBroglie Equation and equation for Planck's Energy R ab=ℏυ(ρ ab). As an energy level of a photon relates to frequency(υ) it can be incorporated into the energy density of matter as can the cosmologic constant which is the energy density of the near vacuum of space.

As $\hbar = h\frac{1}{2\pi}$ the equation becomes $\mathbb{R}(n)abcd = \prod \frac{n^{+1\pi\omega}}{n^{\pi\omega'}} + 2\pi(R\ abc\ +\ or\ -\frac{1}{2}R\ g\frac{ab}{h\nu\rho}ab$ as the Energy Density of matter times Planck's Constant is in the denominator and the 2π goes into the numerator as ℏ becomes $h\frac{1}{2\neq}$. The result is the equation of a Circle Cicumference = 2πR where the circumference is space – time and the Radius is the sumtotal of all Riemann Forces of Nature. This is the compactified (curled up) form of at least type IIA Closed String Theory and perhaps all five String Theories which are mutually dual to each other form M-Theory which is Nature. The permutations or sub-equations to this relationship are enormous totaling at least 524,000 and if inclusive of the spiral operator acting on angular momentum ω may be in the trillions or more. As a conclusion the relationship space – time = $\frac{space}{times}$ times a constant of ~ $\frac{1}{c^2}$ has virtually all equations of Nature as permutations which are 131,076(4)=~524,288 and may or may not be all the equations of Nature. Please note that the denominator ρ ab = R abℏ and ρab incorporates Λ ba which incorporate the energy density in a false vacuum of space. As I ℏ describes the wave function in the Schrodinger Equation the denominator must include i so

$$\mathbb{R}(n)abcd = \prod_0^\infty 2^n + 1\pi\omega i \to \frac{j}{2^{n\pi\omega} j} \to i\ 2\pi\left[R\ a\ b\ c + or\ -\frac{1}{2}R\ g\ ab\right] \otimes i\hbar[\Lambda ba + \rho\ ab]$$

where Λ ba is incorporated into ρ ab as the energy density of matter and h = Planck's Constant. $i = \sqrt{-1}$. The compactification as a circle is circumference = space-time the radius is the sum of all Riemann Forces of Nature and the diameter is the sumtotal of all positive and negative Riemann Forces of Nature. Circumference=2πR of is string theory (type IIA) or M – Theory which is the total of all five string theories type I, II, IIA, Heterotic 8x8 and the SO(32). The spiral operator of the infinite product along 0 to infinity eigenstates of the expanding component of the component sphere or $2^{(n+1)}\pi$ I → j and the contracting component of circles from that of infinite diameter as in flat space – time and no curvature to a point with infinite curvature which is also the intersection of three Planes or $2^{\wedge}n\pi$ where the radius of the expanding and contracting circles forming the cone which is asymptotic to a spiral increasing or decreasing in diameter and radius. In this case the radius from the initial to the final event (i→j) and the radius from the final to the initial event(j→i) comprises an infinite number of constants in nature over an infinite number of eigenstates of energy and as the infinite product of the denominator goes to infinity the entire expression of the spiral operator goes to zero or vanishes from The Equation of Everything leaving Circumference (Curved Lorenzian or Riemannian Space-time)=2π(total Riemannian Forces of Nature)as $\hbar$ = h/2π and i$\hbar$ → ψ(r,t) or the wave function of point particle r at time t.

5.1

A Short Treatise on Spacetime at or near Black Hole Event Horizons

By Dr. Mitchell Albert Wick

Based on the Equation of Everything the formula in terms of metric tensors is as follows $\mathbb{R}abcd(n) = \Pi n = \prod_{1\,to\,\infty} \frac{1}{2}{}^n \pi R\infty R\,abc - \frac{1}{2R}g\,ab\frac{1}{\rho}$ ab in other words the Riemann Metric of all Riemann Forces in n dimensional space Is the spiral operator with the infinite momentum limit R ∞ in the part of the operator which is 2πR representing the circumference of a sphere of nea???

Infinite diameter and the circumference shrinking from infinite diameter for spacetime to a point of infinite curvature. The infinite diameter component has flat spacetime and the point infinite curvature spacetime. In]a black hole the event horizon has nearly 0(zero) spacetime in the conical which is asymptotic to the spiral component of spacetime as indicated by the spiral operator acting on the equation of everything and with the infinite momentum limit R(infinity) when operating on the right side of The Equation of Everything we get the infinite product

Dr. Mitchell Wick

from 0 to n dimensions of the reciprocal of 2 πR ∞ which becomes ∞ so its reciprocal becomes 0 making the right side of the equation 0 While the left side of the equation representing Riemannian Spacetime in the n dimensional state also 0 (zero) at the infinite momentum Limit R ∞ which is the case at the event horizon of a black hole. Therefore with a near infinite momentum Momentum due to extremely dense mass and extreme gravity or the near infinite curvature of spacetime to a point spacetime becomes constricted to a very small area with the time component dilated to nearly infinity. Again this Equation of Everything reads as follow

$$\mathbb{R}(n)abcd = \Pi(n=1 \text{ to } \infty \frac{1}{2^{n\pi R\infty}} \text{ or } \frac{1}{2^{n\pi}} R\infty^{-1} (R \; abc - \frac{1}{2R} g \frac{ab}{\rho} ab$$

where ρ = energy density of matter with the metric g ab and R∞ replacing g ab in the spiral operator.

5.1 THE STACKING OF UNIVERSES IN THE MULTIVERSE

As one notes that a confluence of slices which are imperfect circles formulates the cone asymptotic to the spiral as components of the space-time continuum. The infinite product in the spiral operator has in the numerator $2^{\wedge}n+1$(pi)operating on the function of angular momentum from the initial to the final state so the circles move from a point with infinite curvature and virtually no space or diameter in the initial state or eigenstate to an infinite diameter circle with zero curvature or flat space-time as in an infinite space vacuum in the final state. The denominator is $2^{\wedge}n$(pi)operating on angular momentum from the final state to the initial state. Based on the concept that $2^{\wedge}n+1>2^{\wedge}n$ the circles in the numerator are ahead of the circles in the denominator indicating a lag between the final state of flat curvature of a of infinite space going to an infinite curvature point and going from an infinite curvature point to an infinite diameter circle of flat space-time. THIS LAG RELATES TO THE FIRST EVENT OR FORMATION OF THE SPACETIME CONTINUUM. The time oscillation paradox forming from an infinite number of parallel planes representing D-0-branes whereby a rogue tachyon or tachyons dropped to below light speed causing the time oscillation with the potential energy of 10^77 joules with leptons and gluons going forward in time while tachyons went backwards in time forming the spin,centrifuge effect,the string dimensions and the earliest universes. As a consequence these multi-verse(universes) would be stacked like pancakes initially prior to their Big Bangs,Big Swirls,Spins in which time expansions and contractions would occur.

5.2 DO GHOSTS REALLY EXIST?

Like the tachyon which has been mathematically proven to exist ghosts are mathematically proven to exist. Tachyons have never been isolated as they turn into bosons when they travel below the speed of light(3x10^8 meters/sec). In the case of ghosts or paranormal phenomenon there is NEW DATA FROM THE HADRON COLLIDER AT CERN, Switzerland that seems to disprove the existence of ghosts. There are articles illustrating this as on a level of strings and membrances (supersmall)slippage between dimensions such as the 11th dimension don't show irregularities or perturbations which should exist if ghosts or paranormal phenomena exist. THIS IS A STAGGERING CONTRADICATION AT 40 % of people believe that ghosts and paranormal phenomena exist and this author's Equation of Everything neither proves nor disproves the existence of ghosts which makes that equation of little value (flocinauctonhibilification). THIS DOES NOT DISPROVE THE EXISTANCE OF GHOSTS but the data from Cern rises a "reasonable doubt" although K-suryon waves with specific energy levels and frequencies can carry ghosts and paranormal phenomena as a carrier or transport wave and this can be mathematically proven(K capture:a physical phenomenon often taught in advanced chemistry texts)which may result from x rays and gamma rays carrying other waves (k)which alter the frequency and energy level of the electromagnetic radiation above the predicted amount. All the evidence of ghosts is soft aor anecdotal while the data from Cern is more substantial. Despite this there still may be measurement error at the Hadron Collider which illustrates min-black holes that last for a millisecond or less and if ghosts or paranormal phenomena exist there should be a clustering around these min-black holes.

CHAPTER 6

THE WAVE PROPERTIES OF MATTER AND SPACETIME BY DR. MITCHELL ALBERT WICK

Particle have wave properties according to the deBroglie Equation $\lambda = \hbar \frac{}{mc}$ where $\hbar$ = Planck's Constant and Space – time acts as a perfect fluid. $D = \frac{c}{R}$ where D = the distance which is a function of variable speed of light resulting from the resistance of space-time acting on a particular paricle where $c = \frac{3 \times 10^8 \, meters}{sec}$. The distance of a phopton propagating an infinite distance in space – time = Energy/–$\hbar$ where $\hbar$ = Planck's Constant for the energy of a photon.

According to the perfect fluid equation and hydrostatics Euler's Equation for Quasilinear Hyperbolic Equation governing Adiabatic Inviscid Flow acts as a conductive operator whereby the incompressible Euler Equation with constant or uniform density is Du/Dt=-∇w + g where the vector product of nabla∇. u = 0 and $\frac{D}{Dt}$ is the material operator in time. In the LaGrangian form u = flow velocity vectors or vector components in N – ∇ dimensional space such that u1, us,... un is the flow velocity vector and the Nablaoaperator shows flow velocity divergence with regard to flow velocity gradients and specific pressure. u. ∇ is the convective operator with w=specific theromodynamic work g=body acceleration per unit area and the Euler Momentum Equation with regard to uniform density is $\frac{\partial u}{\partial t} + u.\nabla u = -\nabla w + g$ where ∇.u = 0. The mechanic pressure flow with uniform density $\rho 0$ = incomprehensible constant. So $\nabla w = \nabla\left(\dfrac{p}{p0}\right) = \dfrac{1}{\rho 0 \nabla p}$ with the flow velocity requiring a solenoid field with the continuity equation $\dfrac{\partial f \partial p}{\partial t} = 0$ with a uniform density varying in time... The math for the Perfect Fluid Equation of space – time acting on a point particle in N – dimensional space is $\dfrac{\partial uj}{\partial t} + \sum i = 1 \text{ tp } N \dfrac{uj \partial (ui + wei)}{\partial ri} wh \dfrac{ere \sum \partial wi}{\partial ri} = 0$ and i and j label N – dimensional space components of space – time. In Einstein in notation $\partial i uj + \partial i(uiuj + w\delta ij) = 0$ where $\partial i wi = 0$ an δ = Kronenecker Delta. In the ground state with zero dimensional variables $u* = \dfrac{u}{u0}$, $r* = \dfrac{r}{r0}$, $t* = \dfrac{u0}{r0t}$ and $p* = \dfrac{w}{u0^2}$, with $\nabla* = r0\nabla$ such that the field unit vector $\dfrac{g = g}{g}$ in the nondimensional form of the Eul form of the Eule???

As was said the non-dimensional form of Euler's Equation with a constant and uniform density is $Du/Dt = -\nabla w + \dfrac{1}{Fr^g}$ where $\nabla . u = 0$ Now for a mass of a point particle of mass m the corresponding wavelength is λ where $\lambda = \hbar \dfrac{}{mc}$ and this applies to the incompressive Euler Equation in N – dimensional space. u = flow velocity vectors relating to relating to wavelength λ. As space – time is a perfect fluid and any and all point particles make waves in an interference pattern si??? Ilar

To to electromagnetic radiation where g is the flow velocity vector that propagates out from the point particle in the fluid of spacetime. $Du/Dt = -\nabla w + g$ w is the specific thermodynamic work = ∇ flow velocity divergence and spacetime pressure gradients. $\dfrac{Du}{Dt} = \lambda$ for the point particle so so in the case of a photon $= \lambda - \nabla w = +g = \hbar \dfrac{}{mc}$ and $-\nabla w + g = h \dfrac{}{mc}$ so for any point particle $-w\nabla + g = \hbar \dfrac{}{mv}$ where v approaches c. Of course $\hbar = 6.63 \times 10^{-34}$ erg seconds. In terms of tensors $u\,ab = \dfrac{1}{R}\,ab$ where R ab is the resistance from inertia against spacetime. Let g ab be the metric of the??? $\dfrac{1}{R}ab \propto UUU$ Ua=g ab/R ab because U ab=1/R ab. g ab.D=C/R substitute R with perfect fluid flowflow velocity vectors. The resistance of spacetime to photons with a nonrusting mass is the inertial moment suggesting the Ricci Tensor. $-\nabla w + g = h\dfrac{}{mc} = \lambda$ and $-\lambda = -(-\nabla w + g)\nabla w - g = -\hbar\dfrac{}{mv}$ where mv approaches the momentum operator p so $D = \dfrac{c}{\nabla w} \cdot g = \dfrac{c}{-}\hbar\dfrac{}{mc} = \dfrac{mc^2}{-}\hbar$. Of course for most point particles approaching v = c E approaches mc² so the distance $D-= \dfrac{mc^2}{-}\hbar$ = energy of the point $\dfrac{particle}{Planck's}$ Constant for a photon energy level so the distance of propagation ??? energy/−ℏ. Doppler waves or ripples appear or occur in space-time with each and every point particle of energy or matter forming a comprehensive interference pattern in space-time which relates to the space-time curvature metric caused by mass where the mass includes point particles and photons. The math previously mentioned indicates E=hv where h is Planck's Constant and is derived from the resistance of space-time by photons using the perfect fluid equation for space-time and using flow velocity vectors. The distance. Using tensors we arrived at U ab=g ab/R ab where U ab is the distance of wave propagation in tensors relating to the metric of the point particle or photon and R ab is the Ricci Tensor relating to inertial mass. It was shown that the inertial mass is equal and opposite to the propagation distance in space-time caused by photons or point particles. It was determined that the distance of propagation=total energy as mc^2/h.

For a point particle of mass approaching end distance subtended as a wavelength in fluid space-time is 1.48×10^51 kg of mass where the mass approaches infinity as c is approached (3×10^8 meters/sec) D=C/R=mc^2/h=1/6.63×10^-34 times(3×10^8)2(mass). This comes to 1.48×19^51 kg of mass. The distance subtended by the propagating wave in fluid space-time is a progressive gradient approaching infinite distance where the grad Grad(mass0/(1-v^2/c^2)^1/2.

Ghosts are generally flat matter and they are represented by electromagnetic radiation of low frequency bands. They sometimes manifest themselves on film as orbs and at time vorticies; assuming of course there is no trick photography. The two dimensional lattice formula as applied to the Ising Equation reflecting electromagnetic radiation may apply in the case of extremely low energy bands. One can provide tomes of anecdoctal information of ghost sightings and other paranormal phenomena and if K-suryon waves do exist with a frequency and energy level for consciousness waves then it is possible that there may be some validity to these anecdotal sightings.

In the two dimensional Ising Model there is a ferromagnetic case which results in a phase transition as shown by the Boltzman Equation. By Griffith's Inequality long range interactions occur while Peierl's Argument proves positive magnetization at low temperatures. Onsager'e Exact Solution shows that the free energy of the Ising Model on an anisotropic square lattice when the magnetic field approaches zero H→ 0 *as the thermodynamic limit as a function of temeperature and the horizonatal and vertical* interaction energies are E1 and E2 causing the equation

$$-\beta f = \ln 2 + 1/8\pi \wedge 2 \int_0^{2\pi} d\theta \int_0^{2\pi} d\theta 2 \ln [\cos h(2\beta E1) \cos h (2\beta 2E2) - \sin h\left(\frac{2E1}{kT}c\right) \sin h\left(\frac{2E2}{kTc}\right).$$

Here f=free energy of the Ising System and the expression k relates to the Boltzman Constant with h relating to the thermodynamic limit while sin h cos h are hyperbolic functions which are asymptotic in nature as they approach 0. Of course with asymptotic functions the temperatures reciprocate with the phase transition resulting in the value of 1. The critical temperature occurs in the isotropic case when the vertical and horizontal interaction energies are equal E1=E2=E and the critical temperature T c occurs at $c = 2E / k \ln\left(1 + \sqrt{2}\right.$ *based on the Boltzmann Equation with k relating to the Boltzman Constant which relates* to phase transition. When the interaction energies E1 and E2 are negative the Ising Model an anti-ferromagnet. As a square magnet is bi-partate, it is invariant in the change when the magnetic field H approaches 0 so the free energy and and critical temperature are the same in the anti-ferromagnetic case. The Ising Model for a long periodic matrix has a partition function $\sum ij\ S\ ijSij + 1 + S\ i, jSi +$

Dr. Mitchell Wick

1. *J WHERE IN THIS CASE i is the direction of space and j is the direction of time.*

This is a path integral with the sum over all the spin histories. This can be re-written as a Hamiltonian with a unitary rotation between time+$\Delta time$ or $U = e^{iH\Delta t}$. *The product of U matricies with the total time evolution operator is the path integral.* $U^N = (e^{iH\Delta t})^N = \int DX e^{iL}$ *where N is the number of time slices. Using hyperspace interspersed between the time* slices ghosts can be inserted as two dimensional pixels or voxels of information which can be converted to images such as vorticies or orbs which can at times be photographed especially with infrared camera. Utilizing the two dimensional lattice formula at cold temperatures these units of information are ENCODED AND INTERSTICED within hyperspace anc an possibly move within the 11th dimension.

CHAPTER 7

HOW APPARITIONS MAY BE ECHOS FROM OTHER SPACE-TIME MANIFOLDS REFLECTING AS MIRROR OR ECHO DIMENSIONS OVERLAPPING WITH OUR SPACETIME MANIFOLD AND DIMENSIONS

When one thinks of an onion with layers, space-time manifolds can also be considered as an onion but as a perfect fluid. The past can overlap with the "present" as the layers of an onion can overlap if an external unbalanced force is applied such as electromagnetism, nuclear decay, fusion or fission. An H bomb or even an A Bomb or lightning strikes(or a supernova or gamma ray burst)can unscrample space-time layers overlapping past figures or events with present figures or events. EMP bursts in "cold spots "may work in a space-time overlap. As a result one can visualize and even communicate with another individual from the past who is deceased. This is similar to the idea of several motion pictures except an aurora borealis wouldn't have sufficient energy or force to affect t he transfer. Progressive dilations and constrictions of time causing progressive dilations and constriction of space by these dilations and constrictions of time can mathematically cause a layering of space-time in it's region of topologic space. These layers can be caused by sudden changes in inertial mass within the region causing the curvature of space-time in the region to COIL and UNCOIL at different rates like water spouts in the perfect fluid of space-time. This layering is like the atmosphere separating between the troposphere, stratosphere and ionosphere. Within these layers echos from other time dilations can act as a diffusive effect or even partial impressions acting as osmosis. Between the layers are Einstein- Padowsky-Rosen Bridges which are totally continuous with only kinetic energy equaling the cosmologic constant separating the layers. Sudden changes in inertial mass required to move layers of coiled space-time such that impressions from the past(or future)can bleed through would occur during a Big Crunch,Supernova,Big Bang,or the formation of a star from nebulae.

7.1 ENTANGLEMENT;HOW FAR DOES IT GO?

Lately,new scientific evidence(Science News;Emily Conover vol.191 #84/29/2017) indicates that at least millions of bits of information on a sub-molecular level become entangled with each other. This has far reaching implementations as entangled particles with data can mix or exchange information of any typeanywhere. Can this be associated with paranormal phenomena?" Mystics" who via séances can purportedly contact the dead may in a few select cases have their frontal and temporal lobes of their brain entangled with data from deceased relatives or other individuals. In my previous book "Mega-physics III; Nothing Doesn't Exist "it was explained by this author that "Spooky Action at a Distance" can show information exchange with quantum entanglement between an evaporating black hole and adjacent black holes or even black holes in other galaxies in milliseconds thus solving "The Hawking Paradox". Can ghost sitiings be related to entanglement? If memories or images from the past are stored in the hippocampus of the brain of an individual can electromagnetic impressions which are two dimensional become encoded in the memories of certain perceptive "receivers" which would subsequently be entangled with bits of information containing the spirit and manifest as a ghost siteing? Are senders and receivers all encompassed in each and every brain along the surface as indicated by the University of Arizona(microtubules which may acts as receivers and senders of K-suryon waves).

CHAPTER 8

HOW CAN TIME TRAVEL BACKWARDS OCCUR WITHOUT INCORPORATING A FORCE EQUIVALENT TO THE BIG BANG OR BIG CRUNCH?

As was stated previously, time travel backwards would appear as if a movie moving in reverse either slowly or quickly depending on the rate of recession of the sequencing of events. Remember also that at the event horizon of any active black hole time dilates and slows or practically stops;practically but not completely as time wraps around and constricts space and space cannot be constricted to infinity making space-time zero or a local space-time (spaceless)vacuum which would form a chain reaction impoding a universe in a Big Flush effect in the ocean of fluid spacetime rather than just a whirlpool. Wormholes or Einstein Padowsky Rosen Bridges are strictly theoretical and have not been empirically proven to exist;however many physicists embrace the idea that wormholes can form at the event horizon of any black hole. If this is true and it's a big if,then the total number of universes as previously mentioned can be equal to n times the number of black holes which exist now and throughout the past(including those that evaporated). If this is true and there are 750 billion galaxies in this universe with 500 million black holes existing now that's $7.5 \times 10^9 (5 \times 10^6)$ black holes+black holes that evaporated $= 37.5 \times 10^{15} = 3.75 \times 10^{16}$ black holes and if each black hole has just one wormhole it can indicate over 3.75×10^{16} universes in the multi-verse or more as evaporated black holes would have to be included and assuming only one wormhole per black hole when there may be more than one. Despite this if a wormhole left our universe at a black hole event horizon where the sequencing of events began with the first event of our universe but was a significantly later event in the space-time continuum(new evidence from the U.K. has indicated that our universe was preceded by at least one other indicating the concept of a Big Crunch with the previous universe led to a Big Bang in ours or that the "Bouncing Universe Theory"holds)then the universe the time traveler would arrive at would be in the PAST if the new universe had a Big Bang,Big Swirl or another relative first event relative to the Big Bang in our universe. Conversely if the destination universe would have had a Big Bang would be before that of our universe then the time traveler would move into the future relative to our universe. Of course space travel in general with high velocities approaching the speed of light would have space-time constrict and time dilate so the space traveler would naturally go into the future relative to the earth. Considering the idea that the Hadron Collider at Cern was able to manufacture mini-black holes perhaps they can also product min-wormholes in which subatomic particles besides tachyons can be transported to other universes where their Big Bang isn't coordinated with ours in space-time and would therefore be transported either into the past or future.

8.2 A TRUE MECHANISM FOR TIME TRAVEL BACKWARDS;

The fact of the matter is this;it takes the power equivalent to "The Big Bang"in reverse as a 'Big Crunch" or 6.75×10^{34} erg of energy. Space-time must constrict around the time vehicle without it constricting inside the time vehicle. This would occur at the event horizon of any active black hole or if the time vehicle achieved the mass of a collapsed universe ;a collapsed galaxy may not produce enough power. The mass of the collapsed universe would be 10^{54}kg as has been previously mentioned. Using the Lorenzian Transformations a mass of 10^{54} kg=initial mass/$(1-v^2/c^2)^{1/2}$ so the velocity can be calculated for the time vehicle of say 2000kg.2000kg is 2×10^3kg and this velocity is calculated as follows:10^{54}kg/$(1- v^2/3 \times 10^8 \text{meters/sec})^{1/2}$. As the velocity of the time vehicle approaches 3×10^8meters/sec the mass of the time vehicle approaches 10^{54}kg. At that instant time is dilated significantly but not at infinite dilation as of yet. Space is constricted by dilated time but not to a point of infinite curvature yet and since space is constructed with leptons and gluons the fabric of constricted space can be disrupted by virtually massless gluons being converted with leptons from 99.9999% potential energy into kinetic energy,heat,gamma and x rays which forms quarks from the leptons and gluons releasing The Strong Force as a Super-strong Force. This may crush the time vehicle. This can only be prevented by a force of equal and opposite from the time vehicle to the surrounding constricting time. Some of these ideas were brought about by this author's book "What is the Dimension of Time?".

8.3 HOW COULD WE COEXIST IN THE SAME SPACE AS 'THE BIG BANG' IF TIME THE DIMENSION DIDN'T EXIST?

This could lead to a complete non-sequeter as even if the Einstein-Padowski-Rosen Bridge did allow travel between sequestered and disparate multiple dimensions almost 100 percent of those dimensions would be Planck Length or smaller formulating a Super-brane. Now assume that "The Big Bang" occurred in a microdimension that was Planck length (which is a boundary in physics)or smaller ;there would still be a bleeding out phenomenon from The Big Bang into other dimensions including the macroscopic dimensions. This bleeding out would be detectable as an increase in temp as the temperature post-Big Bang reached billions of degrees kelvin. Now we all know that the thermometers on earth that are centigrade or kelvin at off by 2.74 degrees kelvin(register 2.74 degrees kelvin or centigrade too cold versus the actual temperature) so to show that the time independent case of the Schrodinger Equation holds to unify Quantum Mechanics and Relativity one must prove that the bleeding out temperature is 2.74 degrees kelvin which is the temperature of the false vacuum of deep space plus the background microwave radiaition. Using Occam's Razor(Ockham's Razor)gives us the answer. Which is more likely? That time exists and all events are sequenced to prevent cohabitation of all events simultaneously in the same space or all events everywhere occurred in the same space most of which are at Planck Length or even smaller. If the ladder was true you could replicate all past events with a powerful enough microscope one that isolates event that at below Planck Length. At our level of technology developing this device to prove the existing of sequencing of events and therefore time hasn't been reached yet. Yet the LIGO projects having detected gravity waves near black holes coalescing implies space-time curvature. As space cannot be curved without a curvaing metric a curving metric must exist and that is mass as the curvature of space-time is caused by mass. But tdoes that mean the curvature of space is caused by mass. Since time constricts and dilates space and since net curvature or reciprocal curvature(flatness)is caused by time which is dilated or constricted then space without time must be flat,totally flat even if at infinite mass. The universe isn't totally flat as gravity waves curved space-time by 5×10^{-29} radians which is more precise than Einstein's approximatation of the mass of this universe(1). Therefore according to Occam's Razor the simplest explanation is that events must be sequenced as coexisting with The Big Bang would surely be measurable even with our present technology at below Planck Length.

8.4 HOW DOES TIME WORK IN RELATION TO SPACE?

The following conclusions may be intuitively obvious but could be difficult to visualize so as the author I will attempt to devise a system which incorporate time, space and the rate of sequencing of events. Time by definition is the sequencing of events. The rate of the sequencing determines if time dilates or is constricted relative to space. If time dilates it constricts space and if time constricts it dilates space. The dilation of space causes a flattening effect on space as in a vacuum and the constriction of space causes what approaches infinite curvature and can constrict space to a point with dilated time curving the space towards infinity. As space-time acts as a perfect fluid it would appear as a progressively decreasing dimple in a whirlpool in an ocean of spacetime in the near vacuum of deep space time is constricted and curvature approaches flatness(no curvature). Constricted time goes infinitely fast and dilated time goes infinitely slow so as a vacuum is approached time speeds up and and one propels into the future. The rate of the sequencing of events increases as time constricts and decreases as time dilates according to the observer time traveler. As time constricts or dilates space the rate of sequencing of events constricts or dilates space. Reversing times arrow reversing the sequencing of events and therefore must reverse the rate of dilation or constricting of space by time(which can be though of as constricting or dilating space like a rope rapping around a ball or a cow). Going backwards in time can be thought of as traveling to another universe whose first event occurs after the first event in our universe;but as the destination location may not be the same as the origin location the time traveler could land in the vacuum of intergalactic space at a temperature of just above absolute zero 5 billion years in the past relative to the time sequencing of our universe. This can occur by utilizing a wormhole or Einstein Padowsky Rosen Bridge near the event horizon of any active black hole or merging black holes such as the mini black holes in the HadronCollider in Cern,Switzerland.

8.4 EXTREMOPHILES CAN LIVE IN THE ATMOSPHERE AND MOLTEN LAVA OF VENUS

Bacteria can adapt to extremes of temperature and radiation as mold has been found in high gamma ray sources and the Hadron Collider at Cern,Switzerland. It is believed that after a cataclysmic planetary collision approximately 5 billion years ago whereby our moon was formed the atmosphere was extremely similar to that of Venus today. No water,mostly carbon dioxide and nitrogen from which extremophic life formed in volcanic lava with temperatures similar to Venus today. This protolife resulted from a millisecond gamma ray burst from the fusion of two stars parsecs away which turned D.NA. and R.N.A in from inert organic compounds into chromosomes which are alive. Bacterial speciies were found to be able to withstand temperatures of 800 degrees farrenheit and radiation similar to a gamma ray burst or extreme x rays. Therefore while the area of the earth, venus and mars are compatible with life the earth is most compatible. Despite this as venus and mars could very well have been incorporated in the same gamma ray burst as the earth extremophiles may very well exist on venus as they did on the earth 3.5 billion years ago. It is also possible that extremophiles may exist under the surface of Mars either in bacterial or viral form.

Dr. Mitchell Wick

Our moon which stabilizes the wobble of the 23.5 degree tilt of the earth's axis may have been formed by a planetary collision 4 to 4.5 billion years ago. This wobble if not stabilized would cause a similar phenomenon as with Mars ;evaporation of the atmosphere ;which is why depletion of the ozone layer from excessive solar flare activity could possibly change the angle of tilt of the earth from 23.5 degrees to 22.5 degrees if the mass increases or 24.5 degrees if the mass decreases of the polar ice cap. This may not be sufficient to cancel out the moon's effect on the poles but could still affect the earth. The moon(our moon) may have been caused by a planetary collision 4 billion years ago. A likely candidate for the collision is Venus which acted like two billiard balls hitting each other on a pool table where Venus was initially a part of the earth(their components at that time were virtually identical) and Venus spun off towards the sun with a trajectory which was altered by angular momentum causing it to orbit the sun in the same plane but in a different direction. Venus's orbit is counter to the orbits of other planet which indicates that it was hurled into that spot of the solar system. Indeed if Jupiter is the sun's binary star rather than a planet; a portion of Jupiter could have been a huge flare like object which impacted the earth causing it to split form Venus(which was thrown off) and the moon which stayed in a stable orbit. If this is true Venus's origins were actually part of the earth and extremophilic actvitiy on Venus is of high probability more than mars.

CHAPTER 9

HOW HEURISTIC ALGORHYTHMS USED FOR ARTIFICAL INTELLIGENCE CAN COMPLETE THE ONTOLOGIC PROOF(MATHEMATICAL PROOF OF GOD AS THE COMILATION OF THE MASSSIVE HIGGS FIELD AND THE TACHYON WHERE THE TACHYON IS THE BRAIN AND THE HIGGS FIELD IS THE BODY.)

The energy required for the Higgs Field and the massive Higgs Boson is 10^{77}joules. The Tensor Virial theorem proves that with metric tensors the massive Higgs Field is a huge vortex which is the logarhythmic compilation of a 216 digit number. This proof was incomplete as it didn't mathematically prove thought but AI(artificial intelligence)is asymptotic to thought and K-suryon waves are consciousness waves with a wavelength,frequency and energy level. The Heuristic Algorhythmics complete the Ontologic Proof of God as they approximate though tas an asymptote. This will show God when there is a matching of tensors and heuristic algorhythms as long as there not mixing mathematics.

The Kalb Raymond Action is the most likely pathway for any metric describing an open or s=closed string. As an action it has the highest probability magnitude for the path integral of the function of the tensors of the first,second third and fourth degree going up to the nth degree as described by the open set which contains the subsets from a1 to n where $n< \infty$ *regarding eigenstates of energy over n dimensions. The statistically probability distribution* Follows the quantum mechanical distribution of any event defined by the metric "g"which is measurable. Heuristic algorhythms follow probability distributions of the likelihood of each and every event with regard to space-time and therefore can easily be used in Quantum Mechanics although "fuzzy logic"can be used for the case of AI or artificial intelligence. Statistics is important as well as q values which involve error measurements or describe regions of error due to measurement error made by the intelligent observer. If the intelligent observer is part of what's being measured the results will be skewed and if the Higgs Field is what's being measured the intelligent observer must in the case of "The God Field"must be the tachyon or permanent semi-radius tachyon which acts as the "brain"of the Higgs Field and is separated from the Higgs Boson by the boundary of the speed of light or "c". Using "fuzzy"logic associated with Heuristic Algorhythms shows a distribution curve with the height of the curve or the maximum height being the Kalb-Raymond Action for the closed or open string;the the tensor virial theorem is applied to apply to the vortex of the Higgs Field which incorporates the vortex of the space-time continuum. Heuritstic algorhythms apply to the permanent semi-radius tachyon as the brain and the thought component of the Higgs Field is based on choices and outcomes based on these choices which have a perfect bell shaped probability distribution. AS THE ACTION FORMULA FOR ANY METRIC 'g"can follow a bell shaped probability distribution which is NOT RANDOM(space-time curvature metric is related directly to the $-g$ or minus metric to the half power)then the action formula follows heuristic algorhythms. The effect of creation from the Time Oscillation Paradox was caused by a "rogue"permanent semi-radius tachyon dropping to the spped of light from over light speed. Was this a chaotic action or was it deterministic. It was deterministic if there was conscious thought impelling the permanent semi-radius tachyon to drop to the speed of light. Were their choices or was this a random event? Based on Heuristic Algorhythms there is a set of probabilities for tachyons to act in a random manner and another set of tachyons which acted out of choice.{permanent semi-radius tachyons}={permanent semiradius tachyons[random],permanent semi-radius tachyons[deterministic or by choice]}. IF ONLY ONE ROGUE TACHYON DROPPED TO THE SPEED OF LIGHT FROM

ABOVE LIGHT SPEED THE PROBABILITY OF THE ACTION BEING DETERMINISTIC IS GREATER THAN BEING RANDOM OR CHAOTIC. IF MANY PERMANENT SEMIRADIUS TACHYONS DROPPED TO THE SPEED OF LIGHT AS A GROUP OR SUBGROUP WHICH WERE INTERRECONNECTED THE PROBABILITY OF RANDOMNESS IS INCREASED. PLEASE NOTE THE SECOND LAW OF THERMODYNAMICS WHICH STATES THAT ENTROPY FOLLOWS WHERE THINGS GO FROM A MORE ORDERED STATE TO A LESS ORDERED STATE OR CHAOTIC STATE. AS GROUPS OR SUBGROUPS OF PERMANENT SEMI-RADIUS TACHYONS ARE LESS ORDERED OR HAVE A GREATER ENTROPY THAN A SINGLE TACHYON THE PROBABILITY OF THIS BEING A CHAOTIC ACTION IS INCREASED. There is no clear way to establish which of these quantum states are correct especially if the groupings of permanent semiradius tachyons are interconnected as in a brain, however this configuration would indicate THOUGHT ACTING ON THE HIGGS FIELD WHICH WOULD INDICATE ORDERED CHAOS(Mega-physics;A New Look at the Universe). The confluence of tachyons can act like cell assemblies(Hebb's Theory of Neurobiotaxis) which act similar to neurons or a neural net with logic subsystems. In this case what may appear to be random or chaotic may in actuality be deterministic.

9.1 DESCRIPTION OF THE HIGGS FIELD WITH INTERACTIONS

Describing "The God Field or Particle "has been difficult in attempting to fuse, science, philosophy and religion. The definition of "catholic"is universal;therefore cathalocism is the religion of the universe. Based oon this there must be new definitions. It is indeed difficult to describe the Higgs Field in totality as some assumptions will be difficult for most of the scientific community to "swallow",however as the Higgs 'Boson or Higgs' Field describes the "body" of "God" which can convert energy into matter and back into energy. The semi-permanent hemi-radius tachyon has always existed and will always exist as we perceive time in terms of relatively constricted time and dilated space(a relative vacuum space) The point of creation which was the Time Oscillation Paradox had a rogue tachyon or tachyons drop to the speed of light which is a relative boundary(soft vs. hard boundary). Note:The only hard boundary is that of spacelessness or nothing;which doesn't exist). Was this drop of the rogue tachyon deterministic or random? If it was a group or grouping of tachyons going backwards in time to encounter leptons and gluons turning into Bosons(which go forward in time);then the event would be deterministic and there would be an argument that tachyons are the brain and nervous system of the Higgs'Field. Incorporating an idea first broached by Veneziano in 1993 that this universe is "alive"(at first though of as a preposterous idea but with Quantum Mechanics not so preposterous)one can describe the Higgs'Field(God?)more accurately. When time dilates it constricts space;when time constricts it dilates space,which is why traveling into an infinite space vacuum propels one into the future. In constricted time,time moves almost infinitely fast into the future(times' arrow forward. In dilated time,time moves almost infinitely slow relative to the observer giving flat curvature to no curvature in infinite flat space and space is constricted by the dilated time into an infinite curvature point or dimple in perfect fluid space-time. Galaxies rotate around a central black hole and rotate very slowly relative to the observer. Therefore they rotate in dilated time relative to us much as the event horizon of any active black hole exists in dilated time and constricted space(Schwarzchild Space-time)relative to the observer... Think of time as wrapping around space and constricting it much as a rope strangles a ball until the ball is constricted to a point if the rope is infinitely long. If the rate of rotation of each and every galaxy is perceived as faster it can be considered a dynamo producing energy much as mitochondria does in a unicellular organism. Black holes can be like pores in skin or an excretory function eliminating "waste material "and "The Big Bang" can be considered like fertilization. If a unicellular organism or a lymphocyte is able to conceive an entire body it would be analogous to the observer conceiving the Higgs'Field. Coupling these lofty ideas with tachyons acting as a brain and spinal cord to the body of the Higgs'Field acquiring energy from rotating galaxies and excreting from black holes one can determine THE MEANING OF LIFE. As a white blood cell in our bodies contributes to the well being of our bodies we would contribute to the well being of the massive Higgs' Field and

Higgs' Boson with tachyons acting as the brain and spinal cord and with "birth "being the" Big Bang" which was not the first event as described earilier. As fertilization would not be the First Event but a birth in a continuum of universes;the multi-verse or a collection of Higgs'Fields with tachyons acting as the brain and spinal cord. This concept clearly fuses religion,philosophy and science.

9.2 ONTOLOGIC PROOF THAT THE HIGGS FIELD WITH HIGGS BOSON AND TACHYONS INCORPORATE THOUGHT USING HEURISTIC ALGORHYTHMS

The Kalb Raymond Action utilizing the Tensor Virial Theorem reveals the ACTION OF THE HIGGS FIELD DESCRIBES THE ACTION OF AN OPEN OR CLOSED STRING.

HEURISTIC ALGORHYTHMS DEVUSE ARTIFICAL INTELLIGENCE (thought with computers). The Higgs Field is the body of "God" and tachyons (permanent semi-radius tachyons) are the brain. K-suryon waves describe consciousnessnsnwaves with frequencies, wavelength and energies... Proof: To get the math of heuristic algorhythms and tensor calculus inserted into a proof of equations proving the Tensor Viriral Theorem has consciousness and thought. The Action Formula $S=-1/2k^2(-g)^{1/2} R$ has g as any metric and -g refers to entropy of the metric g. As entropy goes from an ordered state to a less ordered state from the Second Law of Thermodynamics "g"=-entropy or a less to a more ordered state and -g=+entropy from a less to a more ordered state. R=space-time curvature variant from the metric "g". The most likely action used "fuzzy logic" and probability curves or its most likely groupings of tachyons causing an ordered state of an infinite number of totally parallel planes and a finite number of permanent semi-radius tachyons to form a vortex of the space-time continuum which is a less ordered state. The action "S" is a range or approximation of the action or the MOST LIKELY ACTION OF THE METRIC "g" curving space-time R. The term -g implies an increase entropy of the metric "g". A confluence of permanent semi-radius tachyons dropping to light speed follows the law of entropy and one "rogue" tachyon droppint to the speed of light also does but to a less degree than the confluence or a neural net of permanent semi-radius tachyons. The Time Oscillation Paradox caused by a confluence of set of tachyons dropping to light speed has a much higher probability distribution and is much more likely than a single rogue tachyon. The confluence of tachyons dropping to light speed would be much much faster in forming the Time Oscillation Paradox than a single (random) rogue tachyon. The Action formula fits probability and Quantum States much better than a single "rogue" tachyon. The peak of the probability curve with a confluence of tachyons and their corresponding Actions is much much higher than the probability curve of a single "rogue" tachyon and its corresponding Action. The equation $\sum_{\epsilon}^{\infty-\epsilon}$ *(S1, S2, S3,... Sn) where S = action over n eigenstates of energy from epsilon (very small amount) to infinity — epsilon. This describes creation with a probability amplitude which is maximal proving* Deterministic "chaos" or thought by the permanent semi-radius tachyons. The deceleration of tachyons to light speed from above light speed comes from the Law of Conservation of Momentum where a transfer of momentum between tachyons causes the components of the momentum to transfer from one tachyon to another as a an interconnect network.

Example: momentum=momentum)x+momentum)y+momentum)z+...+mom entum)n. As the tachyons communicate with each other as electron spin paring does with Spooky Action at a Distance the components of the velocities of the tachyons change while the total velocity remains constant. One velocity component would slow down to light speed while another component would increase the same or similar amount or anything inbetween above light speed. Couppling this with the Interconnectiveness Theory of Bell and Spooky Action at a Distance defines thought in terms of math utilizing "fuzzy "logic and Heuristic Algorhythms.

The Kalb Raymond Action is the most likely pathway for any metric describing an open or s=closed string. As an action it has the highest probability magnitude for the path integral of the function of the tensors of the first, second third and fourth degree going up to the nth degree as described by the open set which contains the subsets from a1 to n where n< ∞ regarding eigenstates of energy over n dimensions. The statistically probability distribution Follows the quantum mechanical distribution of any event defined by the metric "g" which is measurable. Heuristic algorhythms follow probability distributions of the likelihood of each and every event with regard to space-time and therefore can easily be used in Quantum Mechanics although "fuzzy logic" can be used for the case of AI or artificial intelligence. Statistics is important as well as q values which involve error measurements or describe regions of error due to measurement error made by the intelligent observer. If the intelligent observer is part of what's being measured the results will be skewed and if the Higgs Field is what's being measured the intelligent observer must in the case of "The God Field" must be the tachyon or permanent semi-radius tachyon which acts as the "brain" of the Higgs Field and is separated from the Higgs Boson by the boundary of the speed of light or "c". Using "fuzzy" logic associated with Heuristic Algorhythms shows a distribution curve with the height of the curve or the maximum height being the Kalb-Raymond Action for the closed or open string; the the tensor virial theorem is applied to apply to the vortex of the Higgs Field which incorporates the vortex of the space-time continuum. Heuritstic algorhythms apply to the permanent semi-radius tachyon as the brain and the thought component of the Higgs Field is based on choices and outcomes based on these choices which have a perfect bell shaped probability distribution. AS THE ACTION FORMULA FOR ANY METRIC 'g" can follow a bell shaped probability distribution which is NOT RANDOM (space-time curvature metric is related directly to the -g or minus metric to the half power) then the action formula follows heuristic algorhythms. The effect of creation from the Time Oscillation Paradox was caused by a "rogue" permanent semi-radius tachyon dropping to the spped of light from over light speed. Was this a chaotic action or was it deterministic. It was deterministic if there was conscious thought impelling the permanent semi-radius tachyon to drop to the speed of light. Were their choices or was this a random event? Based on Heuristic Algorhythms there is a set of probabilities for tachyons to act in a random manner and another set of tachyons which acted out of choice. {permanent semi-radius tachyons}={permanent semiradius tachyons[random], permanent semi-radius tachyons [deterministic or by choice]}. IF ONLY ONE ROGUE TACHYON DROPPED TO THE SPEED OF LIGHT FROM ABOVE LIGHT SPEED THE PROBABILITY OF THE ACTION BEING DETERMINISTIC IS GREATER THAN BEING RANDOM OR CHAOTIC. IF MANY PERMANENT SEMI-RADIUS TACHYONS DROPPED TO THE SPEED OF LIGHT AS A GROUP OR SUBGROUP WHICH WERE INTERCONNECTED THE PROBABILITY OF RANDOMNESS IS INCREASED. PLEASE NOTE THE SECOND LAW OF THERMODYNAMICS WHICH STATES THAT ENTROPY FOLLOWS WHERE THINGS GO FROM A MORE ORDERED STATE TO A LESS ORDERED STATE OR CHAOTIC STATE. AS GROUPS OR SUBGROUPS OF PERMANENT SEMI-RADIUS TACHYONS ARE LESS ORDERED OR HAVE A GREATER ENTROPY THAN A SINGLE TACHYON THE PROBABILITY OF THIS BEING A CHAOTIC ACTION IS INCREASED. There is no clear way to establish which of these quantum states are correct especially if the groupings of permanent semiradius tachyons are interconnected as in a brain, however this configuration would indicate THOUGHT ACTING ON THE HIGGS FIELD WHICH WOULD INDICATE ORDERED CHAOS(Mega-physics;A New Look at the Universe). The confluence of tachyons can act like cell assembies(Hebb's Theory of Neurobiotaxis) which act similar to neurons or a neural net with logic subsystems. In this case what may appear to be random or chaotic may in actuality be deterministic.

9.3 DESCRIPTION OF THE HIGGS FIELD WITH INTERACTIONS

Describing "The God Field or Particle" has been difficult in attempting to fuse, science, philosophy and religion. The definition of "catholic" is universal;therefore cathalocism is the religion of the universe. Based oon this there must be new definitions. It is indeed difficult to describe the Higgs Field in totality as some assumptions will be difficult for most of the scientific community to "swallow", however as the Higgs 'Boson or Higgs' Field describes the "body" of "God" which can convert energy into matter and back into energy. The semi-permanent hemi-radius tachyon has always existed and will always exist as we perceive time in terms of relatively constricted time and dilated space(a relative vacuum space) The point of creation which was the Time Oscillation Paradox had a rogue tachyon or tachyons drop to the speed of light which is a relative boundary(soft vs. hard boundary). Note:The only hard boundary is that of spacelessness or nothing;which doesn't exist). Was this drop of the rogue tachyon deterministic or random? If it was a group or grouping of tachyons going backwards in time to encounter leptons and gluons turning into Bosons(which go forward in time);then the event would be deterministic and there would be an argument that tachyons are the brain and nervous system of the Higgs'Field. Incorporating an idea first broached by Veneziano in 1993 that this universe is "alive"(at first though of as a preposterous idea but with Quantum Mechanics not so preposterous)one can describe the Higgs'Field(God?)more accurately. When time dilates it constricts space;when time constricts it dilates space,which is why traveling into an infinite space vacuum propels one into the future. In constricted time,time moves almost infinitely fast into the future(times' arrow forward. In dilated time,time moves almost infinitely slow relative to the observer giving flat curvature to no curvature in infinite flat space and space is constricted by the dilated time into an infinite curvature point or dimple in perfect fluid space-time. Galaxies rotate around a central black hole and rotate very slowly relative to the observer. Therefore they rotate in dilated time relative to us much as the event horizon of any active black hole exists in dilated time and constricted space(Schwarzchild Space-time)relative to the observer... Think of time as wrapping around space and constricting it much as a rope strangles a ball until the ball is constricted to a point if the rope is infinitely long. If the rate of rotation of each and every galaxy is perceived as faster it can be considered a dynamo producing energy much as mitochondria does in a unicellular organism. Black holes can be like pores in skin or an excretory function eliminating "waste material "and "The Big Bang" can be considered like fertilization. If a unicellular organism or a lymphocyte is able to conceive an entire body it would be analogous to the observer conceiving the Higgs'Field. Coupling these lofty ideas with tachyons acting as a brain and spinal cord to the body of the Higgs'Field acquiring energy from rotating galaxies and excreting from black holes one can determine THE MEANING OF LIFE. As a white blood cell in our bodies contributes to the well being of our bodies we would contribute to the well being of the massive Higgs' Field and Higgs'Boson with tachyons acting as the brain and spinal cord and with "birth "being the" Big Bang" which was not the first event as described earilier. As fertilization would not be the First Event but a birth in a continuum of universes;the multi-verse or a collection of Higgs'Fields with tachyons acting as the brain and spinal cord. This concept clearly fuses religion,philosophy and science.

9.4 ONTOLOGIC PROOF THAT THE HIGGS FIELD WITH HIGGS BOSON AND TACHYONS INCORPORATE THOUGHT USING HEURISTIC ALGORHYTHMS

The Kalb Raymond Action utilizing the Tensor Virial Theorem reveals the ACTION OF THE HIGGS FIELD DESCRIBES THE ACTION OF AN OPEN OR CLOSED STRING.

HEURISTIC ALGORHYTHMS DEVUSE ARTIFICAL INTELLIGENCE (thought with computers).

The Higgs Field is the body of "God" and tachyons (permanent semi-radius tachyons) are the brain. K-suryon waves describe consciousnessnsnwaves with frequencies, wavelength and energies...

Proof: To get the math of heuristic algorhythms and tensor calculus inserted into a proof of equations proving the Tensor Viriral Theorem has consciousness and thought. The Action Formula $S=-1/2k^2(-g)^{1/2}R$ has g as any metric and -g refers to entropy of the metric g. As entropy goes from an ordered state to a less ordered state from the Second Law of Thermodynamics "g"=-entropy or a less to a more ordered state and -g=+entropy from a less to a more ordered state. R=space-time curvature variant from the metric "g". The most likely action used "fuzzy logic" and probability curves or its most likely groupings of tachyons causing an ordered state of an infinite number of totally parallel planes and a finite number of permanent semi-radius tachyons to form a vortex of the space-time continuum which is a less ordered state. The action "S" is a range or approximation of the action or the MOST LIKELY ACTION OF THE METRIC "g" curving space-time R. The term -g implies an increase entropy of the metric "g". A confluence of permanent semi-radius tachyons dropping to light speed follows the law of entropy and one "rogue" tachyon droppint to the speed of light also does but to a less degree than the confluence or a neural net of permanent semi-radius tachyons. The Time Oscillation Paradox caused by a confluence of set of tachyons dropping to light speed has a much higher probability distribution and is much more likely than a single rogue tachyon. The confluence of tachyons dropping to light speed would be much much faster in forming the Time Oscillation Paradox than a single (random) rogue tachyon. The Action formula fits probability and Quantum States much better than a single "rogue" tachyon. The peak of the probability curve with a confluence of tachyons and their corresponding Actions is much much higher than the probability curve of a single "rogue" tachyon and its corresponding Action. The equation $\sum_{\epsilon}^{\infty-\epsilon}$ *(S1, S2, S3,... Sn)* where *S = action over n eigenstates of energy from epsilon (very small amount) to infinity — epsilon. This describes creation with a probability amplitude which is maximal proving* Deterministic "chaos" or thought by the permanent semi-radius tachyons. The deceleration of tachyons to light speed from above light speed comes from the Law of Conservation of Momentum where a transfer of momentum between tachyons causes the components of the momentum to transfer from one tachyon to another as a an interconnect network. Example:

momentum=momentum)x+momentum)y+momentum)z+...+momentum)n. As the tachyons communicate with each other as electron spin paring does with Spooky Action at a Distance the components of the velocities of the tachyons change while the total velocity remains constant. One velocity component would slow down to light speed while another component would increase the same or similar amount or anything inbetween above light speed. Couppling this with the Interconnectiveness Theory of Bell and Spooky Action at a Distance defines thought in terms of math utilizing "fuzzy "logic and Heuristic Algorhythms.

Dr. Mitchell Wick

The Kalb Raymond Action is the most likely pathway for any metric describing an open or s=closed string. As an action it has the highest probability magnitude for the path integral of the function of the tensors of the first, second third and fourth degree going up to the nth degree as described by the open set which contains the subsets from a1 to n where n< ∞ *regarding eigenstates of energy over n dimensions. The statistically probability distribution* Follows the quantum mechanical distribution of any event defined by the metric "g" which is measurable. Heuristic algorhythms follow probability distributions of the likelihood of each and every event with regard to space-time and therefore can easily be used in Quantum Mechanics although "fuzzy logic" can be used for the case of AI or artificial intelligence. Statistics is important as well as q values which involve error measurements or describe regions of error due to measurement error made by the intelligent observer. If the intelligent observer is part of what's being measured the results will be skewed and if the Higgs Field is what's being measured the intelligent observer must in the case of "The God Field" must be the tachyon or permanent semi-radius tachyon which acts as the "brain" of the Higgs Field and is separated from the Higgs Boson by the boundary of the speed of light or "c". Using "fuzzy" logic associated with Heuristic Algorhythms shows a distribution curve with the height of the curve or the maximum height being the Kalb-Raymond Action for the closed or open string; the the tensor virial theorem is applied to apply to the vortex of the Higgs Field which incorporates the vortex of the space-time continuum. Heuritstic algorhythms apply to the permanent semi-radius tachyon as the brain and the thought component of the Higgs Field is based on choices and outcomes based on these choices which have a perfect bell shaped probability distribution. AS THE ACTION FORMULA FOR ANY METRIC 'g" can follow a bell shaped probability distribution which is NOT RANDOM (space-time curvature metric is related directly to the -g or minus metric to the half power) then the action formula follows heuristic algorhythms. The effect of creation from the Time Oscillation Paradox was caused by a "rogue" permanent semi-radius tachyon dropping to the spped of light from over light speed. Was this a chaotic action or was it deterministic. It was deterministic if there was conscious thought impelling the permanent semi-radius tachyon to drop to the speed of light. Were their choices or was this a random event? Based on Heuristic Algorhythms there is a set of probabilities for tachyons to act in a random manner and another set of tachyons which acted out of choice. {permanent semi-radius tachyons}={permanent semiradius tachyons[random], permanent semi-radius tachyons [deterministic or by choice]}. IF ONLY ONE ROGUE TACHYON DROPPED TO THE SPEED OF LIGHT FROM ABOVE LIGHT SPEED THE PROBABILITY OF THE ACTION BEING DETERMINISTIC IS GREATER THAN BEING RANDOM OR CHAOTIC. IF MANY PERMANENT SEMI-RADIUS TACHYONS DROPPED TO THE SPEED OF LIGHT AS A GROUP OR SUBGROUP WHICH WERE INTERRECONNECTED THE PROBABILITY OF RANDOMNESS IS INCREASED. PLEASE NOTE THE SECOND LAW OF THERMODYNAMICS WHICH STATES THAT ENTROPY FOLLOWS WHERE THINGS GO FROM A MORE ORDERED STATE TO A LESS ORDERED STATE OR CHAOTIC STATE. AS GROUPS OR SUBGROUPS OF PERMANENT SEMI-RADIUS TACHYONS ARE LESS ORDERED OR HAVE A GREATER ENTROPY THAN A SINGLE TACHYON THE PROBABILITY OF THIS BEING A CHAOTIC ACTION IS INCREASED. There is no clear way to establish which of these quantum states are correct especially if the groupings of permanent semiradius tachyons are interconnected as in a brain, however this configuration would indicate THOUGHT ACTING ON THE HIGGS FIELD WHICH WOULD INDICATE ORDERED CHAOS(Mega-physics;A New Look at the Universe). The confluence of tachyons can act like cell assembies(Hebb's Theory of Neurobiotaxis) which act similar to neurons or a neural net with logic subsystems. In this case what may appear to be random or chaotic may in actuality be deterministic.

THE KALB RAYMOND ACTION RELATES TO THE HIGGS FIELD AND BECOMES THE ACTION OF A STRING(type I,II,IIa) OPEN OR CLOSED STRING UTILIZING THE TENSOR VIRIAL THEOREM TO SHOW THE HIGGS FIELD WITH THE TACHYON ACTING AS A "BRAIN' IS THE 'GOD' FIELD OR THE HIGGS BOSON RELATES TO THE BODY OF THE'GOD PARTICLE'

Quantifying thought can be accomplished using heuristic algorhythms which are also used in AI(Artifical Intelligence)in computers. Of course K-suryon waves which have a characteristic frequency,wavelength and energy level are cosndiered to be consciousness waves.

HOW CAN THE TACHYON BEING THE "BRAIN" OF THE HIGGS FIELD WITH DETERMINISTIC THOUGHT INVOLVED IN THE POINT OF CREATION?

The tachyon has the ability to travel back in time or reverse "Times'Arrow". It was previously noted by this author that it takes the power equivalent to "The Big Bang" as an implosion to reverse Times'Arrow and the tachyon particle(which cannot be isolated and is only proven mathematically)has this ability. It requires 10^{77} joules to reverse "The Big Bang" and reverse Times Arrow either in the environment with the tachyon still going forward in time or for the tachyon to go backward in time. Therefore the tachyon has the ability to generate power equivalent to The Big Bang or Big Crunch as 10^{77} joules. This power generation can be performed by the "God Particle" and it was postulated previously that the tachyon acts as a "brain" for the Higgs Field which acts as a body.

CHAPTER 10

HOW DID LIFE INITIALLY BEGIN?

BY DR. MITCHELL ALBERT WICK

ABSTRACT

Approximately 3.5 billion years ago the first chromosomes on the planet earth formed which was the blueprint for rudimentary life. Prior to this there was messenger RNA and transfer RNA as well as DNA(desoxyribonucleic acid). DNA,mRNA, and tRNA are inert(non-living) but chromosomes are living. What was the catatlyst that reschuffled the DNA and RNA into chromosome bodies? 3.5 billion years ago the atmosphere around the earth contained methane CH_4,hydrogen sulfite H_2S,carbon dioxide CO_2,oxygen(O_2),nitrogen(N_2) and water (H_2O). This was not the atmosphere that surrounds the earth in 2017. Also if there was an ozone layer it was significantly smaller than today, The surface temperature of the earth was closer to venus than earth perhaps 350 degrees farrenheit. There as a catalyst. Approximately 3 billion years ago there was a gamma ray burst from the fusion of two neutron stars which cut through this noxious atmosphere like a knife through butter lasting $1-3 \times 10^{-3}$ sec. This gamma ray burst reschuffled the RNA and DNA breaking it into fragments moving them around like a deck of cards being schuffled and in this case there was gravity effects(curvature of spacetime due to mass)which forced an oscillation,spin and resassembly of the DNA into the formation of 46 chromosomes... The sugars from lipids and carbohydrates developed as well as more complex proteins after the first rudimentary life formed.

Dr. Mitchell Wick

HOW TO SEARCH FOR LIFE ELSEWHERE

Locate regions of gamma ray burst in the past and present then find celestial bodies which have water,oxygen and carbon. This undertaking may be difficult as finding the locations of Gamma Ray Bursts of very short duration with our technology may determine locations in the present but finding them from millions of years ago may be very difficult unless a footprint or signature of the gamma ray burst can be found. The size of gamma ray bursts can be very small or as large as a solar system.

10.1 HOW CAN MASS CREATE SPACE IF SPACE CANNOT BE CREATED OR DESTROYED ACCORDING TO THE LAW OF CONSERVATION OF DIMENSIONS?

According to Albert Einstein mass creates space around it. This can only be possible if masslessness is impossible. Spacelessness is impossible. Even in the primordial vacuum or Fermionic State there was energy due to the Law of Conservation of Energy which stated that energy cannot be created or destroyed(the total energy of a system). In my previous book "Megaphysics III;Nothing Doesn't Exist" this author stated that space is comprised of Fermions which are almost massless and contain almost infinite potential energy with a miniscule amount of kinetic energy to keep the infinite number of totally parallel planes from contacting each other in the zero dimensional state as in D-O-Branes. These "massless"fermions which comprise space are leptons which form the fabric of space and gluons to hold the fabric of space together. While these fermions are almost massless they do have mass and while they comprise and hold space together it doesn't contradict Einstein. Another assumption is that a container doesn't require contents to exist. In the case of the infinite parallel planes the container is comprised of nearly massless Fermions and exist without touching each other. (please note again that the cosmologic constant of nature incorporates the kinetic energy to keep the planes apart). So a container isn't nothing and doesn't contain nothing but still exists;so that assertion is also true

10.2 HYBRID MASS ANTI-MATTER AND 'GHOSTS'

As one may speculate with anti-matter and hybrid mass the mass of dark matter=mass of ordinary matter/i and hybrid mass is an oscillation of positive matter and negative matter causing an oscillation in space-time during that frequent. This oscillation effect can correspond to a nexus junction in space-time as discussed in my first book "Megaphysics, A New Look at the Universe"(2003)which is a crossing of dimensions. Space-time of course acts as a perfect fluid so this nexus junction or overlap may appear as a "dimple"in the pool of space-time where the mass of odd dimensions always manifests itself and the mass associated with even dimensions only manifests itself intermittently. Due to the fact that dark matter seems to curve space-time inward displaying gravity effects more than is recorded it seems to oppose the effects of dark energy and the cosmologic constant which flatten space-time or push it outward with anti-gravity. A nexus junction could be a portal in space-time that is only Planck Length 10^{-33} cm or may be larger however due to the oscillation effect larger nexus junctions may cause too much disruption of space-time which can cause a local "paradox" There must be an exact alignment between the angle of the dimensions and the curvature of space caused by time in order for the junction to hold and display "occurrences". The frequency of these exact alignments are rare and infrequent and are difficult to record or document. It has also been mathematically proven by this author that as anti-particles repel each other with anti-gravity(dark energy emanates from there) it would be difficult for apparitions to maintain cohesion very long(perhaps 5 to 10 seconds)and therefore if ghosts are hybrid mass they would appear only at nexus junctions intermittent with the hybrid mass.

It was certainly clear that Quantum Mechanics and Relativity were unified when the wavefunction in the time independent case as in the event horizon of any active black hole is reached where space-time is maximally constricted. As the cosmologic Constant or energy density of a vacuum is very close to 0 or 10^{-53} it can be considered episilon or as an approximatation can be called approaching zero as an asymptote depending on how much flattening occurs with space-mass due to the

Dr. Mitchell Wick

absence of mass besides the Background Microwave Radiation fro "The Big Bang". There is naturally reciprocal curvature or any unwinding of space-time (dilation) as the dimension of time unwinds toward zero curvature or flatness. This generally occurs in the false vacuum state and can be considered in the approximation $\frac{\partial \Psi}{dt} =\sim \Gamma[\rho+0]$ or $\frac{\partial \Psi}{\partial t} \sim \Gamma[\det \rho(r,t)]$

$$\frac{\partial \Psi(r,t)}{\partial t} =\sim \Gamma \, \rho(r,t)$$

In this case the determinant is 2×2 determinant for a point particle(r) with respect to time or a tensor of the second degree between space-time and mass. The fluctuations of space(constrictions and dilations are incorporated in the vector field acting on the point particle(r) formulating the wave function based on the frequency and energy levels based on Planck's Constant which are the energy levels of a photon or electromagnetic radiation of negligible mass which significantly increases as space-time constricts and time dilates as at the event horizons of any active black hole. Any scattering of this electromagnetic radiation is quashed by the cosstriction of space-time as a current or whirlpool acting like a toilet flush effect. As mass and energy are interchangeable and mass causes the curvature of space-time this equation is RELATIVIZED or has Relativistic Variance. Indeed it is an inch long like the concept behind the Equation of Everything but it is a mathematical construct rather than a verbal description. To prove it incorporates string theory and M theory it must compactify to a circle. The energy density of matter in constricting space-time as in a black hole goes from a circle of near infinite diameter and flat spacetime (no curvature) to a point approximately Planck Length of infinite curvature. These phenomena are curled up and compactified to a point which is a circle of infinite curvature where the energy density of matter is so constricted by spacetime it is like a wound up spring with extreme POTENTIAL ENERGY "while the wave function of a point particle is the unwound state of an unsprung flattened out spring. So as a consequence the concentrated energy density of matter at the event horizon of a black hole becomes more diffuse in nature as it is unwound into flat space-time of the false vacuum of space with kinetic energy and heat reflecting as Hawking Radiation or the quasar effect. In the case of the wave function the geometry is a spiral of gradually increasing diameter from the point particle of infinite curvature with constricted space-time at the event horizon. To put it simply the winding of the spring or the decrease in diameter and increase in curvature of space-time is relativized and the increase in diameter and the decreased curvature or flattening of space-time follows the wave function of the Schroodinger Equation of Quantum Mechanics. There is a continuous set of circles forming a cone of Schwarzchild Space-time and diameter so these all compactify to a circle and described M Theory,type IIA String Theory and due to Wick Rotation the heterotic 8x8 string theory.

HOW OBJECTIVISM TIES IN WITH THE EQUATION OF EVERYTHING AND THE UNFIED FIELD THEORY

The principal tenet of Objectivism is the following :a=a which is the Transitivity Postulate. The transitivity postulate is a very powerful tool in mathematics ;and Ian Rand's philosophy which ties in with "every man for himself" also ties in with the Equation of Everything. The Equation of Everything came from the idea that space-time=space-time as every quantitiy equals itself. Then as time dilates near heavy masses or slows down constricting space in that Region of space-time then it follows that space-time decreases as mass increases and time dilates as dilated time constricts space with progressively increased curvature toward infinite curve of time constricting space to a point. This occurs adjacient to near infinite mass which explains how space-time is inversely proportional to mass. Space-time is directly proportional to space due to the Line Element $ds^2=dx^2+dy^2+dz^2-c^2dt^2+dr^2$ and time as the sequencing of events is the space-time curvature metric known as the effect of gravity/inertial mass causing the curvature. This results in the Equation of Everything space-time is directly proportional to space and inversely proportional to mass as was mentioned in all of this author's previous texts. What's interesting abaout this is it leads to the unification of Quantum Mechanics and Relativity as curved

Riemannian or Lorenzian Space-time(R abcd(n)=the infinite product over n eigenstates of energy of 2^n+1(pi)angular momentum from the initial to the final event/2^n(pi)angular momentum from the final to the initial event[the spiral operator operating on the function of anular momentum with the increasing size of the diameter of the spiral or cone in the numerator and decreasing diameter in the denominator forming an infinite number of constants added to R abc(Euclidian flat space)+ or − 1/2R g ab which is gravity for-1/2R g ab and anti-gravity for +1/2R g ab divided by inertial mass of R ab which is h(bar)times rho ab where h=Planck's Constant and rho is the energy density of matter from Poissons Equation. To include the heterotic principle of flipping between dimensions as in the heterotic 8x8 string theory the square root of negative 1 is multiplied b with h/2(pi)rho. One can Wick Rotate the denominator between dimensions. Note also that ih(bar) times the wave function of a point particle with respect to time in Schrodinger's Equation equals the ih(bar0the energy density of matter of Poisson's Equation plus the Cosmologic Constant. As Poisson's Equation and the energy density of matter can be relativized with respect to curved space-time in the numerator of the Equation of Everything and the fact fact that the Schrodinger Equation is the hallmark of Quantum Mechanics ;they are unified. Another concept which may or may not be assumed also follows from a=a; Everything cannot be created or destroyed. Also everything always existed and always will exist in one form of energy or matter or another considering the number of dimensions discussed. This also follow from the idea that nothing doesn't exist and Energy cannot be created or destroyed. In addition the zero energy state doesnot exist. Of course as mentioned in prior books the Equation of Everything mathematically compactifies to a circle not a sphere proving it to be string theory which is what Michio Kaku was the Equation of Everything. Of course the 524,288 permutations describing the number of equations in nature is added to an infinite number of constants which actually sum up to zero and therefore drop out of the equation.

NATURE FOLLOW OBJECTIVISM OR EVERY MAN FOR HIMSELF

For those who accept evolution whose" tent is survival of the fittest"with adjustments in D.N.A. m-R.N.A and t-R.N.A. one can understand while Objectivism fits in with nature and this as well as other universes in the multiverse."Every person for himself or herself"means doing what's best for that system whether it be alive or inert. As the second Law of Thermodynamics regarding entropy(disorder)reveals that energy must be placed into a system to maintain this ordered state rather than it skewing into disorder. Objectivism show order and disorder in "The Fountainhead" where Ellsworth Tewey is the disordering of the ordered Howard Roark. Tewey preached self sacrifice and quashed individual creativity in exchange for uniformity and "taking orders"even at the expense of one's life. Howard Roark believed an idea was the property of the individual who created the idea and that he had the right to create it or destroy it if not to his satisfaction/seeking perfection and order while Tewey sought chaos and disorder... Nature follows evolutionary ideas to adapt and order chaos while the forces of nature work as obstacles to trigger evolution.

CHAPTER 11

THE FINAL UNIFICATION EQUATION OF QUANTUM MECHANICS AND RELATIVITY

By Dr. Mitchell Albert Wick

$\Psi(r,t) = \Gamma[(\rho(r,t) + \Lambda(r,t)]$

THE WAVE FUNCTION OF A POINT PARTICLE EQUALS THE DERIVATIVE OF THE TENSOR SUM OF THE ENERGY DENSITY OF MATTER WITH RESPECT TO SPACETIME AND MASS FOR POINT PARTICLE r WITH RESPECT TO VARIABLE TIME. ACCORDING TO THE C,P,T THEOREM -t and t are interchangeable WITH RESPECT TO THE ENERGY DENSITY OF MATTER AND THE ENERGY DENSITY OF A VACUUM. THE ENERGY DENSITY OF MATTER AND THE COSMOLOGIC CONSTANT FOLLOW RELATIVISTIC VARIANCE AND THE ENERGY DENSITY OF MATTER RELATES TO THE DUAL VECTOR FIELD OF SPACETIME AND MASS ACCORDING TO POISSON:S EQUATION WITH A 78 5 POISSON PROBABILITY. THE GEOMETRY OF THE WAVE FUNCTION IS THAT OF A STANDING WAVE OSCILLATING BETWEEN y=sin x and y=cos x and any photon scattering is quashed by constricting space-time as photons approach the event horizon of any active black hole as in the time independent portion of Schrodinger's Equation where $\Psi(r) = \rho(r) + \Lambda$

The equation
$$\Re(n)abcd = \prod(n=1 \text{ to } \infty) 2^n + 1\pi\omega \; i \to \frac{j}{2^n \pi} \omega \; j \to i + R \; abc + or - \frac{1}{2R} gab$$
or $-or - i\hbar(\rho \; ab + \Lambda \; ba)$ shows that The Schrodinger Equation
$i\hbar \dfrac{\partial \Psi}{\partial t} = i\hbar(\Gamma \left[\rho(r,t) + \Lambda(r,t)\right] \; i\hbar \dfrac{\partial \Psi}{\partial t} = i\hbar(\rho + \Lambda)$ or for the point particle
r at time t $i\hbar \dfrac{\partial \Psi(r,t)}{\partial t} = i\hbar[\rho(r,t) + \Pi(t,r)]$ Here the derivative operator of a dual vector field of point particle r is the metric g interacting with the region of space-time(t) in Poisson's Equation such that the energy density of matter ρ and the cosmologic constant Λ *which is the energy density of a vacuum are relativized.*

The energy density of matter is ρ *in* $\nabla(r,t) = 4\pi\rho(r,t)$ where $\int_0^\infty \Psi \; dt =$ *the integral of the wave function of the point particle from t =* ∞ *to t =* 0. *Which is from the Schrodinger Equation. As the cosmologic constant represents antigravity the tensors are opposite* to the energy density of matter. Q.E.D.7//13/2017 by Dr. Mitchell Albert Wick

CHAPTER 12

THE RELATIONSHIPS OF BLACK HOLES AND THE NUMBER OF EXTANT UNIVERSES

IN THE MULTIVERSE

There is new information between the two dimensional holographic manifold at the event horizon of any active black hole and the Einstein Padowsky Rosen bridge which may indeed be a wormhole between universes. As we know space-time constricts toward zero due to infinitely dilated time wraps around fluid like space as the event horizon of a black hole is approached. As the gravity of a black hole isn't infinite the curvature of space-time as the event horizon isn't infinite but is constricted as in Schwarzchild Space-time. Based on the symmetry of Schwarzchild Space-time and the extremely condensed area within a black hole;there may very well be an exit and expansion of Schawrzchild Space-time at the opposite end of the throat of the Einstein Padowsky Rosen Bridge or a wormhole. This exit may be another universe which would also solve the Hawking Paradox as information would be spumed out the mouth from the throat of the wormhole o=at the other universe. At this point in time,technology does not know how many universes exist in the multi-verse but based on logic whether there be 2 billion universes or 10^{10^8} universes as computers at Stanford predict one can see an exact mathematical relationship between the number of universes and the number of active black holes in our universe. Whether this relationship is 1:1 or larger is based on whether more than one stable wormhole can emerge from the event horizon of any active black hole. As space-time constricts toward zero as the event horizon is approached this becomes an asymptotic function where zero space-time would indicate ONE WORMHOLE AND A 1:1 realtionship while miniscule space-time would indicate there is room for more than one wormhole. As an example if space-time is constricted by 80 percent at the event horizon the number of wormholes that can fit would involved the diameter and square area of the wormhole (which would have to take a spiral configuration such as a photon trail would take). Despite this as long as there is finite space with finite time wrapped about it the constriction must be finite and the number of wormholes which would fit would be finite but approaching infinity. However if the space was infinite wrapped about by infinitely dilated time one and ONLY ONE WORMHOLE WOULD BE ABLE TO FIT IT INFINITELY CONSTRICTED SPACE-TIME WHICH IS LIKE A PINPOINT PINCH IN FLUID SPACE-TIME. SO THE NUMBER OF UNIVERSES =THE NUMBER OF WORMHOLES AT THE EVENT HORIZON OF ANY ACTIVE BLACK HOLES IN OUR UNIVERSE TIMES THE DEGREE OF SPACE-TIME CONSTRICTION CAUSED BY THE MASS OF THE BLACK HOLE WHICH IS A PATH INTEGRAL RELATING TO SCHWARZCHILD SPACE-TIME. AS a conclusion as the mass of any active black hole varies with the size of the neutron star,galaxy or universe that collapsed(as in the supermassive black hole at the site of "The Big Bang'. This can be calculated as a range of black hole measurements in our 750 billion galaxies with regard to mass and converted into a path integral with regard to space-time and its curvature metric calculated from the action formula where the metric is space-time being curved by the inertial mass of the black hole. As a consequence of this the number of wormholes would be finite in our universe but a huge figure. As an example if there are 10 billion black holes in our galaxy and 750 billion galaxies we ger $10(10^9)(7.5\times10^{11})$ where $10(10^9)$ is the number of black holes in our galaxy and 7.5×10^{11} galaxies in our universe assuming the collapse of stars as a rate is approximately the same in each galaxy. This number must be integrated over 0 to infinite space-time as the limits for the function of the cross sectional area of any representative black hole times the inertial mass of the black hole. Based on this there would be $7.5\times10^{20}\int_0^\infty \frac{du}{u}$ where du/u is the spiral fractal formula acting on the spiral operator operating on space-time.

12.1 THE SUPERBRANE BY DR. MITCHELL ALBERT WICK

A SUPERBRANE is a membrane that encompasses everything and is so small that it encompasses all other membranes and all mass. Space-time can be subdivided down to 10^{-33} cm or Planck Length as the orbifold but below that the nature of space-time is up to speculation. It has been proposed by this author that space-time below Planck Length is formed by a confluence of quantum dots which have an approaching infinite curvature as a point would. These quantum dots vibrate and oscillate into each other making the "nothing space" between them approach zero as the frqency of vibration and oscillation exceed "c" or the speed of light 3×10^8 meters/second which further encompasses the idea that "nothing doesn't exist". It has been postulated by Einstein that the mass of every body exceeds that of 1000 suns. This would normally seem unlikely as space-time curvature in the universe is miniscule at 5×10^{-29} meters whereby most of the spiral curvature would be adjacient to the event horizons of black holes and whereby the majority of this universe is nearly empty space approximately 10^{22} km with a temperature of 2.74 degrees kelvin the BMR from The Big Bang. As 99.99999999999999999% of the universe is empty space without mass or gravity(the curvature of space-time caused by mass)a vast majority of the universe is totally flat or 5×10^{-29} meters of curvature which is asymptotic flatness. However most of the spacetime curvature is so closely adjacent to the mass of any and all particles including superstrings that it makes up the quantum dot matrix where each dot has nearly infinite curvature and dots composed space-time on a level which is smaller than Planck Length. The summation of the curvature of each quantum dot relates directly to the space-time curvature metric of gravity or R g ab as in the action formula $S=-1/2k^2(-g)^{1/2}R$ and develops an almost infinite number for the confluence of quantum dots which oscillate and vibrate at greater than "c" in essence πc or the velocity of spacetime. These quantum dots are so miniscule that they could be 10^{10}

Cm smaller than Planck Length and are therefore virtually unmeasurable with our level of technology but they would comprise the SUPERBRANE which is flexible and encompanies everything with every mass having internal motion which is greater than light speed in vibration and oscillation, The curvature of each component quantum dot approaches that of infinity as a quantum dot is asymptotic to a point and the sum of each quantum dots curvature added to the essentially flat space-time of virtually empty massless space would derive the line element equation or space-time curvature metric on Mintkowski flat space-time as in the equation of everything. And as a multitude of infinite curvature points in spacetime is reduced in size from essentially flat space-time the spiral operator used to measure space-time as the event horizon of a black hole can be utilized for all space approaching this massive superbrane. So the equation $\mathbb{R}abcd = \prod n = 1$ to $\infty \frac{1}{2^{\pi r}} g\, ab\, Rabc - \frac{1}{2R} g \frac{ab}{\rho} a\, b$ where *the infinite or near infinite number of dimensions that are so miniscule*

That they dwarf Planck Length as in the subdivided on second of arc where each plane intersects with an infinite number of other planes as the curvature metric causes the planes to be nonparallel causing an infinite number of intersections or dimensions that are far smaller than Planck Length and incorporated in this MASSIVE SUPERBRANE WHICH EMCOMPASSES EVERYTHING AND HAS APPROACHING INFINITE space-time curvature to reflect the huge density of what apprears and is measured as a limited mass in small objects.

CHAPTER 13

WHAT ARE THE CHANCES OF A PARADOX IN SPACE-TIME IN THE FUTURE?

Our universes expansion is geometric as it approaches older universes in the multiverse which are primarily black holes and it approaches the region of spacetime in the vortex of the space-time continuum as the recoil of the spring effect described in "Megaphysics III;Nothing Doesn't Exist". We know the time dilates as it approaches the extreme mass of the event horizon of any active black hole which means it slows down. Nealry infinitely dilating time constricts space toward a point of Planck Length(10^{-33}cm) which would appear as a dimple in the ocean or Bose=sea of fluid space-time. In the near vacuum of space time speeds up as it constricts toward zero and has a negligible effect or almost no effect on the space of space-time as space-time is flat in a vacuum. As time the dimension not the invention(manually devised clocks)moves almost infinitely fast in a near vacuum due to the lack of mass to curve space-time our time the dimension is moving very quickly into the future which is what makes time travel forward so much easier than time travel into the past(see book "What is the Dimension of Time?" and Megaphysics III; Nothing Doesn't Exist". If time was moving infinitely fast into the future then time travel backwards would be impossible or that which is beyond the comprehension of man to factor in quantum states. Time can move backwards with tachyons which are a fundamental part of the "God Particle" or Higgs Field and the comprehension of the tachyon while mathematically proven may be almost impossible to isolate and it's properties may be beyond the comprehension of man. Despite this with time moving almost infinitely fast into the future the tachyon still moves into the past;something mankind can't seem to do. Once the tachyon moves infinitely fast into the past or infinitely slow into the past it can cause another Time Oscillation Paradox which will causes a dimensional reschuffling of space-time in the space-time continuum. If infinitely fast into the past a paradox of Time Oscillation could occur at any moment,but if nearly infinitely slow backwards it could occur much much later(can't define what much much later is if time the dimension is moving almost infinitely fast into the future. So if the expansion of space-time is increasing geometrically toward the center of the original space-time continuum as a spring effect and if time is moving almost infinitely fast into the future then our universe will have a Big Crunch into a quantum bubble of 10^{-24}cm before Heat Death occurs from a deceleration of the expansion. When this will occur as time isn't traveling infinitely fast but nearly infinitely fast is questionable but as long as the acceleration of galaxies and spacetime is geometrically increasing it will occur sometime in the future. The only thing that will slow down time is nealy infinite mass everywhere. Space-time has negligible mass comprised of leptons and gluons which have negligible mass (Megaphysics III;Nothing Doesn't Exist. and space-time is infinitely large(proven in Megaphysics II:An Explanation of Nature".(infinity times epsilon"very small quantitiy) is still infinity which describes an infinite mass for space-time even when time is constricted toward zero or has almost no constricting effect on space. This mass works against the speeding up effect of time towards a Big Crunch although another Time Oscillation Paradox can still occur although Heat Death(Frank Tipler)is unlikely as the expansion is accelerating.

CHAPTER 14

WHAT IS HYPERSPACE?

Hyperspace is a continuum of points which are the intersection of three planes the xy,xz, yz where each intersection defines a point. There are an infinite number of points which are discontinuous and therefore hyperspace is discontinuous IF THE POINTS ARE STATIC; but as nothing is static as absolute zero is non-breachable these points must have motion if they contain energy (matter included)and at the very least are vibrating as long as a vacuum isn't spaceless(which is impossible). Therefore as long as there is motion hyperspace can be almost continuous as the limit of the distance between each point approaches zero. Therefore it is possible that some paranormal phenomena may be right in front of us, between us and even inside us or inside solid objects and be difficult to detect if they are almost massless like photons or any low energy electromagnetic radiation such as radio waves are if non-corporeal energy traverses hyperspace as it is discontinuous unless blurred or smeared, it is possible that thoughts as well as paranormal "phenomena" traverse hyperspace. As this might be true thoughts can be disseminated through space-time via hyperspace thereby traveling at velocities greater than "c" or 3×10^8 meters/sec. If this is true thoughts have a pathway from the microtubules of the brain in all animal life throughout space-time to other microtubules explaining telepathy as well as K-suryon waves. Even "Spooky Action at a Distance may traverse through this medium. Paranormal phenomena may have their information trapped between the point of the discontinuous hyperspace. Generally things which are normally unexplainable might be explained.

CAN HYPERSPACE BE MATHEMATICALLY DETERMINED WITHIN SPACE-TIME?

Without digressing the Harmonic Spherical Bessel Functions can be applied to space-time in a decreasing diameter from infinity toward zero as a cone asymptotic to a spiral after the first event. Spherical Harmonics utilizing the Legendre Functions and Polynomials of a degree j and order m reveal the equation $(1-z^2)d^2w/dz^2 - 2z\, dw/dz + [j(j+1) - m^2/1-z^2]w = 0$ whereby j and m are complex numbers in format a+or − bi and a+bi=0 and a−bi=0.j $j(z) = (\pi/2z)^{1/2} \otimes J\,j + 1/2(z)$ and $h\,j\,(z) = (\pi/2z)^{1/2} \otimes N\,j + 1/2(z)$ $h\,j(1)(z) = (\pi/2z)^{1/2} \otimes H\,j + 1/2(1)(z)$ $h2(z) = (\pi/2z)^{1/2} \otimes H\,j + 1/$

2/2(z). Here and thence forward $\otimes$ is dot product o or just listed without symbols. As the wave function in the Schrodinger Equation incorporates ih or −ih spherical harmonics ties into space-time with regard to the wave function as indicated in the denominator of the Equation of Everything as + or *−ih(Λ ba + ρ ab) applying it to the toal energy density of matter.* Of course the compactification of string theory is a circle and this is a two dimensional representation of a sphere as a two dimensional hyper-surface being progressive shrunk from an infinite diameter to a point with infinite curvature as a cone asymptotoci to spiral space-time after the first event. Asymptotic expansion of the cylindrical (after time oscillation paradox forming the vortex) and spherical Bessel Function for |z| gives $$Jm(z) = \sqrt{\frac{2}{\pi z\left[Am(z)\cos\left(z - \frac{m\pi}{2} - \frac{\pi}{4}\right) = Bm(z)\sin\left(z - \frac{m\pi}{2} - \frac{\pi}{4}\right)\right]}}$$ and

$$Nm(z) = (2/\pi z)^{1/2}\left[Am(z)\sin\left(z - \frac{m\pi}{2} - \frac{\pi}{4}\right) + B\,m(z)\cos\left(z - \frac{m\pi}{2} - \frac{\pi}{4}\right)\right]$$

with the | arg |z < π(*arg is argument or maximum argument for z where | | is the absolute value*) in radians with respect to z and the expansion component of the spiral operator of space-time acting on angular momentum. The Spherical Bessel Functions relate to space-time as −1/2 e −i n cot theta whereby the Spherical Bessel Functions relate to j $j(z) = (\pi/2z)^{1/2}\, J + 1/2(z)$ and $h\,j\,(z) = (\pi/2z)1/2\, N\,j + 1/2(z)$ $h\,j(1)(z) = (\pi/2z)\, H\,j + 1/2(1)(z)$ *and finally* $h2(z) = (\pi/2z)^{1/2} H(2)j + 1/2(z)$

where by d^2w/dz^2+2/z dw/dz)[1-j(j+1)/z^2]w=0 and j are integral values.; This clearly shows that + or -1/2 e −i n cot theta where theta is z is e-i n cos

$$\frac{z/z}{\sin z \, \frac{z}{z}} \text{ were the } z\text{'s or angles of trajectories as} \neq \text{radians cancel}.$$

Utilizing these spherical Bessel Functions for h0(1)(z)=-e^iz/z and h-1(0)(z)=e^-iz/z one can derive the cotangent of the trajectory of the Big Bang as

cos z/sin z where z=π radians where $j0(z) = \sin z - j - 1(z) = \cos z / z$. Here h0^(2)(z)=ie^-iz/z and h-1^(2)(z)=e^-iz/z such that j i(z)=z^i(-1/z d/dx)^I sin z/z and n i(z)=(-1)^i+1 j-1-1(z) where j=(1,2,3,...). This leaves the n dimensional case of spiral space-time being acted upon as a Spherical Harmonic Bessel Function.

The Spherical Harmonic Bessel Function and the Cylindrical Bessel Function both describe space-time as the Spherical Harmonic Bessel Function describes space-time as a whole and the Asymptotic Cylindrical Bessel Function describes space-time curvature describing space-time as + or -1/2 e − incotangent z or theta in the quantum mechanics description of the equation of everything (see book The Equation of Everything). The Asymptotic Cylindrical Bessel Function describes hyperspace as it is discontinuous blurring or smearing into a continuous curve as per the path integral of $\int_0^\infty du/u = \ln u$ which describes spiral space-time which is a confluent of increasing or decreasing spheres as described by the Spherical Harmonic Bessel Function and the cylindrical effect is asymptotic to the corkscrew configuration of space-time as described by Albert Einstein in 1912. Combining the Spherical Harmonic Bessel Function and the Cylindrical (corkscrew configuration of space-time) Bessel Function produce the spiral configuration of space-time as the spherical harmonic function reduces or increases the diameter of the corkscrew or cylindric configuration of space-time.

CHAPTER 15

THE BIG BOUNCE

Dr. Neil Turok as the Director of the Perimeter Institute of Canada and Dr. Steffen Gielen of the Royal Academy of Physics in London4 stated that with the laws of Quantum Mechanics predominating at Planck Time or 10^{-43} seconds there was a deceleration of the near infinitely dense quantum bubble resulting in a near complete reversal of the Big Bang accelerating the initial Big Bang back towards the center of the vortex of space-time prior to the main expansion of the Big Bang. This point has been debated by physicists since 1922 as there was a question with this initial Big Crunch as to whether or not it returned to a nearly infinitely dense quantum bubble. As this was PRIOR TO THE FORMATION OF HYDROGEN OR HELIUM the world of strings predominated. Using a new model for Conformal Symmetry in the quantum bubble there is extant mathematical proof in models purported by Dr. Steffen Gielen and Dr. Neil Turok. There are apparently new models supporting the idea of a Big Crunch before the Big Bang which was purported by this author in the book "What is the Dimension of Time? "as well as "Megaphysics, A New Look at the Universe".

The concept of conformal symmetry would explain this author's 0,0 point which would be the fracture point in the center of the quantum bubble.

Dr. Mitchell Wick

15.2 QUANTUM GRAVITY

In terms of the Equation of Everything using Planck Mass and the inertial mass R ab we get Curved Lorenzian or Riemannian Space-time as described by Weyl's Conformal Tensor is $C\ abcd(n) = 2\wedge n+1\frac{\pi\omega}{2^{n\pi\omega}} + 2\pi(R\ abc + or - \frac{1}{2}Rg\frac{ab}{\rho}ab$ where $i\hbar(\rho +\Lambda\ ba) = R\ ab$ and $R\ ab = \sqrt{h\frac{c}{8\pi G}}$.

The effect of gravity on Planck's Mass is R g ab which when coupled with the anti-gravity of Planck's mass is the reciprocal of Planck's Mass. It is therefore $(8\pi G/\hbar c)^{\wedge}1/2$ *as the square root in the numerator and denominator equals anything to the zero p???*

Therefore the gravity and antigravity effects by Planck's Mass cancel out as ~0 gravity whereby Weyl's Conformal Tensor or the curved Riemannian tensor for curved space-time=flat or Mintkowski Space-time as in the quantum bubble before the Big Bang.

General Relativity is non-renormalizable with a negative n-dimensional gravitational coupling constant(k). The equivalence principle states that the laws of physics in a gravitational field are identical to those of a local accelerated frame of reference. This explains the relationship of inertial mass and gravity which is the curvature of space-time caused by inertial mass remain a constant regardless of any dimensional space discussed. In the string n-dimensional state space-time is curved in the same way by mass(even string sized) as when a black hole that is supermassive curves space-time as the event horizon is approached. For the equations of motion the power expanding the metric tensor encompassing the solution g(0) μυ(*the Lorenzian Metric*) *of the equations of motion we get the expression* $g\mu\upsilon(x) = g(0)\mu\upsilon + kh\ \mu\upsilon$ where $h\mu\upsilon$ = *graviton field* and $k = \sqrt{Gn}$.

In this case the Graviton Field is the action(s) of inertial mass indicated by R ab or the Ricci Tensor curving flat Minkowski Space-time into

Lorenzian Curved Space-time in a conformal manner as described by Weyl's Conformal Tensor of gravity or C abcd.

The LeGrangian Action

$2\partial\rho\ hv\sigma\partial\ \wedge vh \wedge \rho\sigma$ or $h^{\rho\sigma^{\upsilon}}$. *The Legrangian or Le Grangian Action is the most likely action of the L???*

Metric acting on the graviton field h with regards to µ *and* υ *whereby* µ *relates to the mass curving space-time with* υ *relating to the velocity of motion of* µ *in the graviton field.*

The total action is the sum of the LeGrangian and gauge symmetry partition functions. $\mathcal{L}\ 0 = -\frac{1}{2\partial}\times h\rho\sigma\ V^{\rho\sigma\mu\upsilon\partial\lambda}\ h\mu\upsilon$ and $V\rho\sigma\mu\upsilon = \frac{1}{2\delta}\rho\upsilon\delta\sigma\upsilon - \frac{1}{4\delta}\rho\sigma\delta\mu\upsilon$ *whereby V acts on a dual action tensor field with regards to the graviton*

Inverting the matrix V ρσµυ *reveals the Final Propagate of the fourth degree covariant tensor. Quantum ???*

be expressed in a formally non-renormalizable form as local symmetry leads to what are called Ward Identities which cancel divergent graphs of quantum or topologic loops. Note according to Ward Identities the higher loop counterterms can only exist if the theory is divergent or infinite. As the counterterms of quantum gravity are formally non-renormizable(as mentioned previously) the divergent quantum loops cancel as part of the Ward Identitites the way Bianchi's Identity works with fourth degree covariant and contra-variant tensor of Riemannian Space-time or Weyl's Conformal Space. Based on these arguments "Arg" quantum gravity is a finite or convergent theory. The normalization coefficient of the wave function in quantum mechanics is related to the probability of the wave function of point particle "x" at time "t" existing in space along the boundaries of integration such that the coefficient of $C = \int_{-\infty}^{\infty} \|(x,t)\wedge 2\| = 1$ or 100% probability distribution of the wave function of point particle x at time t is satisfied. Local symmetry groups relate to gauge symmetry in Yang Mills Theory and relates to non-Abelian subgrouping which are finite in nature and relate to Quantum Field Theory and therefore Quantum Gravity. Graviton fields must also follow local gauge symmetry in order to be non-divergent quantum loops and the probability of existence must be 100% in order to avoid "ghosts" or divergent functions. In this author's previous book "What is the Dimension of Time?" it was shown mathematically that trying to prove the existent of nothing reveals only mathematical "ghosts".

The final propagate $\delta\ \mu\rho\delta\upsilon\sigma + \delta\mu\sigma\delta\upsilon\rho - \delta\mu\upsilon\delta\rho\sigma \div k^2 + i\epsilon$ *is the inversion of matrix V $\rho\sigma\mu\nu$* forming quantum gravity in the non-re-normalizable form. Here k=gravitational coupling constant $i = \sqrt{-1}$ and epsilon relates to the Gauss-Bonnet Identity as related to the total derivatives of the product of two anti-symmetric constant tensors whereby the topological invariant implies the square of the curvature tensors of space-time rather than the conventional meaning of a small metric approaching zero. The total number of loops that are quantum(topologic) loops that show relativistic invariance that describe 1-loop counter-terms are finite while the asymptotic or infinite counter-terms which are invariant cancel again with regard to higher quantum loops. {R^2 μυρσ, R μυ^2, R^2} *are counterterms that are invariant whereby 1/R^2 describes space-time curvature of of the metric g μυ*. If Tμυ or the stress energy tensor of the metric gμυ = 0 *then* $R\mu\upsilon - \frac{1}{2}\ g\mu\upsilon R = 0$

Where R μυ represents interial *mass and gμυ represents the metric curving space-time as the effect of gravity. ith regard to one finite quantum loop the general relat??? Time is $R^{2\mu\upsilon\rho\sigma}$. In other words Einstein's Equation of Relativistic gravity with regard* is finite(convergent) with regard to the 1 loop level.

The 2 loop level cannot be canceled by the equations of motion and C μυαβ with regard to quantum gravity is less divergent and Weyl's Conformal Tensors which comprise the Riemann Metric of space-time tensors whereby 1/ϵ reveals divergences which are eliminated by the Gauss – *Bonnet Identity with regard to Quantum Field Theory.*

To conclude, quantum gravity in the lower loops tend to converge into the equivalent of a finite theory but there are still divergent systems associated with larger loops which do not all cancel with the Gauss-Bonnet Identity. In the lower loops the convergence might tend toward

$$\mathcal{L}\, 0 = 1/4[-(\partial v h \rho \sigma)^2 + (\partial \mu h \rho^\rho)^2 - 2\partial \sigma\, h\rho^{\rho\mu} h^{\sigma\mu} +$$

a limit of zero as an asymptote as the stress energy tensor in Einstein's Equation of Motion would tend toward zero in the above mentioned case. Infinite mass is a divergent value as total mass is finite albeit huge which would constrict space-time toward zero with infinite curvature without having space-time converge to zero as this to is an asymptote. The point is that absolute mass-lessness will not be reached even prior to the formation of the quantum bubble or the bubbles of the multi-verse as the components of space had huge potential energy and miniscule kinetic energy separating the infinite number of parallel planes prior to the time oscillation paradox and that anti-gravity as dictated by the cosmologic constant was a factor in separating these planes or D-0-branes from forming the first dimensions prior to the time oscillation paradox. Zero gravity relates to mass -less- ness and does not occur. Quantum anti-gravity from anti-matter strings self repelling cancel out virtually all of the quantum gravity in the quantum bubble and cause the unopposed expansion with the spring effect in the current of space-time from the center of the vortex toward a 0 acceleration point followed by a 180 degree reversal caused by the pull of space—time upon space-time back toward the center of the vortex. Of course this would mean that the weak force of Dark Energy caused by the Cosmologic Constant with antimatter repulsion would be almost exhausted when the spring is released the quantum gravity would actually exceed quantum anti-gravity as the directional moment would change by π *radians or* 180 *degrees. The mass of the quantum bubble is finite and expands in thi*to 10^64 kg. Naturally The Equation of Everything would say that space-time constricts considerably in the quantum bubble

relative to space-time outside the quantum bubble like the center of a whirlpool of perfect fluid space-time but at Planck Time 10^-43 seconds space-time would be flattening from a near infinite curvature of the quantum bubble to curvature which approaches flatness from the Cosmologic Constant and Dark Energy prior to it's shifting it's direction by 180 degrees back toward the center. Of course based on this the time of expansion before the reversal back toward the center of the vortex is based on the magnitude of Dark Energy and it's pushing effect against conformal time and if the accelerated expansion of the galaxies slows toward zero and Heat Death this reversal may not have occurred yet after 13.7 billion years which would precede a Big Crunch when this universe spiral back toward the center of the vortex. It is also possible the reversal occurred less than a second after the Big Bang so the first second expansion of 6.75×10^{34} erg could have been from matter-anti-matter annihilation and repulsion of antiparticles sequester in one pole of the bubble while matter sequestered at the opposite pole. Still this reversal did occur or will occur as Newton's Second Law, The Law of Conservation of energy and a specific location in space-time for the Big Bang to occur would then be localized as well as the center of mass of our universe or conversely of the multi-verse. Finally the formation of the multi-verse can also be postulated in an analogous way as our universe except one quantum bubble would become a 700 googlplex or more quantum bubbles or manifolds of different configurations such as Calabi_-Yau Manifolds and myriad other configurations and in this case the bubble for our universe would follow a Big Crunch of another universe (as previously mentioned). In a sense Quantum Gravity would tend to be convergent as space-time curves inward with gravity (curvature inward of space-time caused by mass)and Quantum Anti-gravity would be reciprocal curvature of the flattening of space-time caused by strange or hybrid mass and might be considered divergent as space-time is curved outward. THIS DOES NOT MEAN THAT DIVERGENT HIGHER LOOPS NOT CANCELLED BY THE GAUSSBONNET IDENTITIES WOULD BE RELATED TO ANTI-GRAVITY AS QUANTUM GRAVITY AND ANTI-GRAVITY ARE OPPSOITE SIDES OF THE SAME COIN;THE WAY SPACE-TIME CURVES. Every question regarding" The Big Bang' calling it a singularity has now been answered. THE LOCATION OF THE BIG BANG IS THE CENTER OF THE SPACE-TIME VORTEX CAUSED BY THE IME OSCILLATION PARADOX ACTING ON THE INFINITE NUMBER OF PARALLEL PLANES. BEFORE THE BIG

BANG WAS NOT NOTHING SO THE LAW OF CONSERVATION OF ENERGY IS PRESERVED AS THERE WAS ALMOST INFINITE POTENTIAL ENERGY OF SPACE IN THE INFINITE PARALLEL PLANES WHICH CONVERTED TO KINETIC ENERGY AT THE TIME OSCILLATION PARADOX. NEWTON'S SECOND LAW HAS THE ACTION OF THE BIG BANG SPRING BACK TOWARD THE CENTER OF THE SPACE-TIME VORTEX AS THE EQUAL BUT OPPPOSITE REACTION. THE QUANTUM BUBBLE OR BUBBLES DEPENDING ON WHETHER THERE IS JUST ONE UNIVERSE OR A MULTI-VERSE DOESN'T EMCOMPASS EVERYTHING AS THE BIG BANG EX NIHILO STATES BECAUSE SPACE ISN'T NOTHING AND SPACE EXISTED PRIOR TO THE BIG BANG. NOTHING IS WITHOUT ANY PROPERTIES AND IS SPACE-LESS, MASS-LESS AND WITHOUT ENERGY EITHER POTENTIAL OR KINETIC. ALSO NOTHING ALWAYS WAS AND WILL ALWAYS BE NOTHING AND NOTHING CANNOT BECOME SOMETHING WITHOUT A CATALYST AND THE CATALYST IS THE ROGUE TACHYON DROPPING TO THE SPEED OF LIGHT TRIGGERING THE TIME OSCILLATION PARADOX AND TACHYONS ARE SOMETHING AS IS SPACE. THEREFORE NOTHING DOESN'T EXIST. RELATIVITY WITH REGARD TO THE MASS DENSITY OF THIS HOMOGENEOUS ISOTROPIC UNIVERSE AND THE BIG BANG

The present mass density of the background microwave radiation at the present is $1-\wedge-3$ or $1/1000$ th of the mass density of matter. The B.M.R. should have been a dominant contribution to the mass density of this universe. As $a \rightarrow 0$ *the temperature in degrees kelvin approach ten million degrees. At this point H approaches H zero or H(0).*

Acceleration factor approaches zero before The Big Bang $.a(\tau) = \dfrac{\frac{4c^1}{4}\tau^1}{2}$ where rho $\rho = \dfrac{3}{32\pi G \tau^2}$ and ρ is the mass density of the universe. Again to simplify $a(\tau) = \sqrt{\tau(4c)}^{-1}\big/4$. The mass density $\rho = n\sum_1^n \alpha\, i\, gi\, \dfrac{\pi^2}{30} \hbar^{3c^5}$ $(KT)^4$ where T is the absolute temperature $K =$ space $-$ time curvature.

Dr. Mitchell Wick

$\alpha i \to \frac{7}{8}$ *for ferminons and 1 for bosons with n being the number of states or eigenstates o???*

With g i=spin degeneration factor and τ = time, Hubble Time or Conformal Time.

If the mass density, KT it acts like zero mass due to 0 space-time curvature which comprises vacuum space comprised of leptons and gluons which are essentially massless with near infinite potential energy. tε' $\frac{a}{a'} = 2\tau$ so as the density = $\sum_1^n \alpha i g i \left(\frac{\pi^2}{30} \hbar^{3c5} \right) KT^4$ so in the denominator of The Equation of Everything the (KT) π(π) ÷ 30ℏ^3c^5)KT^4 as the sum of α i g i is the mass density of this universe. Becomes zero because the space-time curvature in a near vacuum state or Fermionic State approaches zero so if K~0 then KT~0 and KT)^4 ~0 so $R\mathbb{R}abcd(0,1) \to$ infinity in the 0 and first eigenstate. Type equation here.2 π(R abc + or $-\frac{1}{2}$ R gab/ih(Λ ba + ρ ab)have Λ ba + ρ ab = the mass density(KT)⁴ where K → 0 so in the o eigenstate pre Big Bang and pre – First Event space space was clearly infinite and time was infinitely dilated or in ess???

In the Infinite Parallel Plates there was zero curvature as in the Fermionic or Vacuum State. Also the temperature in degrees kelvin would be negligible so the denominator would still approach zero and the numerator would become space- time=space with almost zero gravity and anti-gravity due to the negligible mass of space from leptons and gluons. In other words space-time=space→ ∞.

CHAPTER 16

WHAT CAME FIRST; THE BIG BANG OR THE FORMATION OF THE MULTI-VERSE?

The Big Bang was preceded by a quantum bubble of approximately 50% matter and 50% anti-matter. How did this precise mix form? The most plausible explanation was a collision of two membranes which comprised of a matter universe with physical laws consistent with matter and an anti-matter universe with laws consistent with anti-matter. Of course when anti-matter and matter collide they form an explosion annihilating a goodly percentage of each but there was an IMPLOSION or a BIG CRUNCH with the commixing or collision of the two membranes associated with the two universes. The Big Crunch resulted when the impact of two universes which distorted or disrupted space-time to the degree that it caused a Big Crunch. There is also the possibility of a three way collision between a matter universe, anti-matter universe and a vacuum universe consisting of only space and energy. In the 700×10^{10} googlplex of universes postulated at Stanford University the likelihood of vacuum universes is high ; and they would force the matter and anti-matter universes INTO the vacuum universe to fill the vacuum and it's related space crunching space-time down to 10^{-24} cm in a mix or anti-matter and matter. This of course rotated and the matter settled at one pole and the antimatter at the opposite pole with the 0,0 point in the center whereby a symmetrical throat was formed and was the site of the Big Bang at 10^{-43} seconds later(Planck Time). Of course if it wasn't for the 50:_50 mix of anti-matter and matter one could have presumed that the quantum bubble was formed after matter and the string dimension were formed. Indeed, matter and anti-matter could have been formed then but NOT IN A 50:50 mix. Although with our present scientific knowledge we may be able to ascertain that the universe does have a center of mass at the site of the Big Bang by the increased rotational vectors of space-time as the region is approached by measuring gravity waves; determining if there was a three way collision of membranes developing into a Big Crunch or a two-way collision of membranes developing into a Big Crunch may be beyond our scientific expertise as the Big Bang ex nihilo(out of nothing)was acknowledged no matter how perverse it appears. Also the idea that this universe has no center of mass when Sir Isaac Newton stated that everything with mass has a center of mass or gravity ;and the universe has a mass of 10^{64} kg. iso-tropism is relative based on the observer's relative position which decries the idea of Quantum Mechanics that when an observer is part of what's being observed the results are skewed by the limits of the system being observed. So basically, the universe may not have an edge but does have an interface with space-time comingling into space-time from other universes in the multi-verse. Space exists outside this universe as space-time and not nothing]. There is soft evidence that $10^{10^{10^{10^7}}}$ universes exist in the multi-verse; if our universe had no center of mass or gravity it would be expanding in all directions from what]?Whether expanding from Inflation or a mutual repulsion of antiparticles and particle-anti-particle annihilation in the quantum bubble the 0,0 point which is the point of maximum rotation the rotation and expansion must be ascertainable and measurable in order for the big bang to hold. Even the Steady-State Theory has to begin with the Infinite Parallel Planes Hypothesis in which case the galaxies and universes formed when matter and anti-matter and all types of strange matter or hybrid matter form as the SECOND EVENT. 5.2 THE LAW OF CONSERVATION OF DIMENSIONS THE LAW OF CONSERVATION OF DIMENSIONS in "Mega-physics,A New Look at the Universe"(2003)stated that the total number of a dimensions in a system must remain a constant. A system is defined as a set or subset which is bounded. In the case of the first event the infinite parallel planes comprised of potential energy from gluons, leptons,and tachyons composed one and only one system ;the zero dimensional system.

After the first event a new system or subsystem was formulated which comprised of matter, energy,and all the dimensions which entail string theory($26 \to \infty$). This results in a different system or subsystem to the zero dimensional state before the event even comprised of D-0-Branes. All the remaining membranes were formed after the first event as potential energy turned into matter and kinetic energy, heat,electromagnetism, the Strong Force,the weak force and either other universes or the quantum bubble were formed. Once this subsystem was formed it was non-Abelian as the components were necessary and although the components of space(gluons,leptons and tachyons)existed before and after the first event. This is because no planes are totally parallel after the first event as spacetime was originally spiral and a vortex with string dimensions gravitating toward the center due to the effect of the spin 2 vector bosons. Therefore the total number of dimensions before the first event=0 and after the first event =1 to ∞ *so the total number of dimensions of each system or subsystem were a constant although aga*again space cannot be created or destroyed.{0} before the first event regarding dimensions and {1,2,3,... 26...∞}*after the first event. The comonents pre and post first event include* {tachyons, fermions}and after the first event were {tachyons,fermions{gluons, leptons},spin 2 vector bosons,The Higgs Boson, quarks, neutrinos and all other inclusive hadrons are all composed of strings,potential energy, kinetic energy,heat,electromagnetism, the Strong Force,the Ultra-strong Force(released with the disruption of gluons in the fabric of space being held by leptons)and Orbifolds which are inclusive of Fock Space, Hilbert Space deSitter Space and anti-deSitter Space. Note that the null set{} is impossible,but{0}is the zero dimensional state containing an infinite number of parallel planes is possible,likely and did occur.

CHAPTER 16.1

HOW DID GRAVITY FORM?

Gravity is the curvature of space-time caused by mass. Before the first event space was infinite and mass was negligible caused by leptons, gluons and tachyons (plus possibly photons). During the first event the spin 2 vector boson was formed from the primordial tachyons and fermions during the time oscillation paradox. The rogue tachyon(s) dropped to the speed of light v=c and then was transformed into a massive Higgs Field ;which is spiral in configuration according to the Tensor Virial Theorem(see book "What is the Dimension of Time? ". This massive Higgs Field perturbed space-time into a spiral configuration during the time oscillation forming a spin and centrifuge effect. During the spin component the Higgs Boson underwent a clockwise and counterclockwise spin or (+ or − spin)for each component of the boson cohabiting the Higgs Field and the components of the spin 2 vector caused by the boson helped form the vortex of the Higgs Field and well as that of space-time form. Mathematically this author first book "Mega-physics, A New Look at the Universe "showed that !" ! = ! $\in \ln u \to \ln 0 = \epsilon$ (*approching zero but not 0 as ε was the kinetic energy that always existed.*) Space was composed of leptons and gluons which were almost 100% potential energy but a very small component was kinetic energy which prevented the infinite parallel planes of space from contacting each other as contact between two parallel planes would have produced a space paradox prior to the time paradox as since time is the sequencing of all events having an event such as a space paradox prior to the first event regarding time couldn't happen. If time is infinitely dilated before and after any event including a space paradox then a space paradox wouldn't be an event. As a space paradox is an event and requires time as a necessary component or prerequisite of the paradox; a space paradox couldn't happen without time.

CHAPTER 17

THE BIG BANG

AS DESCRIBED IN THIS AUTHOR'S PREVIOUS BOOK "MEGAPHYSICS II: AN EXPLANATION OF NATURE" THERE WAS A ROTATIONAL COMPONENT TO THE QUANTUM BUBBLE WHEREBY ANTI-MATTER ROTATED IN ONE DIRECTION AND MATTER ROTATED IN THE OPPOSITE DIRECTION WITH THE ROTATIONAL VECTORS GRAVITATING TO THE OPPOSING POLES AS IN A CENTRIFUGE WHICH WAS 10^{-24} cm. With the rotation of matter and anti-matter being 180 degrees from each other and gravitating toward the poles the fracture point with the symmetrical throat which resulted was what this author called the 0,0 point as per the first book "Mega-physics, A New Look at the Universe". As antiparticles express anti-gravity or repulsion to other anti-particles the mutually repulsive force gravitated to he center due to the extreme rotation in opposite directions of matter and anti-matter at the poles. The combination of the mutually repulsive force forming Dark Energy with its anti-gravity and the annihilation of most anti-matter with matter CAUSED THE BIG BANG(Mega-physics II: An Explanation of Nature)and resulted 6.75×10^{34} erg during the first second of The Big Bang. This of course explained the paucity of anti-matter, the cause of the Big Bang and what Dark Energy is. Space-time was original spiral immediately after the first event; the Time Paradox oscillation but after the Big Bang there was a combination of rotation and expansion of space-time simultanteously with the rotational vector diminishing while the expansion vector was increasing. As a consequence any center of mass for this universe will show up with an increasing rotation in space-time curvature with gravity wave measurements as in the LIGO project(see What is the Dimension of Time?")This would approach the site of The Big Bang and the light of this supermassive black hole is such as distance from the Hubble Telescope 10^{10} light years that as a black hole the equipment for detecting an absence of light may lack the sensitivity to detect it even if the light from that site has reached us and can be measured. Even at 10^5 light years which is 100,000years for the light to reach the Hubble IT IS THE ABSENCE OF LIGHT AT THE EVENT HORIZON OF THE SUPERMASSIVE BLACK HOLE AT THE SITE OF THE BIG BANG THAT NEEDS TO BE MEASURED BUT CAN'T AS THAT LIGHT IS COMPLETELY ABSORBED INTO THE BLACK HOLE AND NOT REFLECTED. EVEN IF THE BLACK HOLE BECAME DRY OR INACTIVE AND DISSOLVED INRO SPACE LIKE A DIFFUSION OF SALT MIXING WITH WATER AS A SUSPENSION OF INFORMATION UNIFORMLY SPREAD THROUGHOUT SPACE,WITH OUR TECHNOLOGY THERE IS A LACK OF CAPACITY TO DETECT AND EVALUATE THIS SUSPENSION OF INFORMATION IN SPACE-TIME. This information would give vital evidence of the birth of this universe and while the WOMP in 1996 gave a general view of the universe with clumping of the Background Microwave Radiation after the Big Bang it's still incomplete,which is why the measurement of gravity waves showing space-time curvature is necessary to determine a double spiral configuration whereby one spiral or vortex is counterclockwise and the other clockwise illustrate the opposing rotations of matter and anti-matter before the orb blast. Of course as these vectors were 180 degrees from each other or *π radians the result was π radians clockwise and π radians counterclockwise as in the Wick* Rotation of {x,y,z,ict} and {-x,-y,-z,ic(-t)} or *π − π radians where −reflects the opposite direction of the spin of antimatter to matter.* The C.P.T.

Theorem(charge, parity,time) indicates that time and −time should be equivalent and with regard to Wick Rotation being heterotic(flipping between different dimensions)space-time which is negative only appears negative with respect to the x,y, and z axes not the −x,-y,-z axes so the resulting 2 π *radians of the orb blast with a trajectory of* π *radians or* 180 *degrees holds.*

CHAPTER 18

WHAT IS THE CONFIGURATION OF THE MULTI-VERSE?

Space-time was originally a vortex after the first event and mathematically spacetime curvature follows the spiral fractal formula according to Einstein in 1912 and this author's first book "Mega-physics,,A New Look at the Universe" and the matter and anti-matter forming after the formation of strings and all Branes above the D-0- Brane progressed to form $10^{10^{10^{10^7}}}$ universes in the spiral space-time continuum with space-time diverging toward infinity in every and all directions except the center of the vortex. This is why space-time travels at $2.2 \times 10^{35}(c)$ meters. sec while in our universe it only travels at πc *acting as a current being pulled by space – time traveling at* 2.2×10^{35}© meters/sec The reason for the geometric acceleration of space-time is the push of Dark Energy and THE PULL OF SPACE-TIME FROM THE PERIPHERY TOWARDS THE CENTER OF THE VORTEX OF SPACE-TIME IN WHICH THE MULTIVERSE EXISTS. IT IS ALSO POSSIBLE THAT A BIG CRUNCH OCCURRED WHICH PULLED SPACE-TIME AT THAT VELOCITY TO FILL THE SUPERSTRONG FORCE OF THE DISRUPTION OF GLUONS ON LEPTONS IN SPACE-TIME. HOWEVER, IT IS MORE LIKELY ACCORDING TO OCCAM'S RAZOR that space-time is being pulled into the center of the spacetime vortex. Recall Occam's Razor states that all things being equal the simplest explanation is generally the correct one and a singularity resulting in a Big Rip, tear or Crunch of space-time is less likely than space-time being pulled AS A PERFECT FLUID INTO THE

CENTER OF THE SPACE-TIME VORTEX. The space-time continuum is spiral regardless of whether there is one universe or a multiverse of 700 googoplex or 10^{10^7} universes or more as the acceleration of space-time as a current radiates towards the center of the vortex and this acceleration is independent of the number of universes involved in the space-time continuum. Math based on the comparative mass of the universe and the multiverse implies one but that might be incomplete as again there is soft evidence of other universes. Acceleration is like a Big Flush of fluid like space-time with extraordinary angular momentum despite negligible mass due to the acceleration acting as a current in a perfect fluid region R(p). This is due to the velocity being $2.2 \times 10^{35}(c)$ meters/sec.

18.2 WHERE IS OUR UNIVERSE'S RELATIVE POSITION IN THE SPACE-TIME CONTINUUM?

18.3 THE RELATIVE POSITION OF OUR UNIVERSE IN THE VORTEX IS BASED ON THE RATIO OF SPACE-TIME IN OUR UNIVERSE TO SPACE-TIME PULLING IT FROM THE REMAINDER OF THE SPACE-TIME CONTINUUM TO THE CENTER OF THE VORTEX.

The velocity ratio is $\dfrac{\pi c}{2.2 \times 10^{35}(c)}$ *which puts our universe in the region of space – time which has flatness equivalent to 5×10^{-29} radiants or the total curvature of our universe.* The flattening effect is from the Cosmologic Constant relating to Dark Energy and the enormous total pull of space-time onto space-time in our universe. The latter has a more powerful flattening effect than the weak force of Dark Energy. This relates to the mutual repulsion of anti-particles which is so sporadic in space due to great distances between each and every anti-particle that it makes the flatness of space-time primarily from space-times acceleration pull. As totally flat space-time is only possible when space is infinite and time infinitely dilated; totally flat spacetime mimics the vacuum state of space with infinitely dilated time and is at the extreme outer periphery in all measurable and immeasurable directions. Despite this, still with an extreme velocity pulling on space-time (which is almost motionless) and the infinite space vacuum state or fermionic state is approached space-time curvature is best measure with our technology to determine the location of our universe in the space-time continuum... Our universe while having extremely diminutive space-time curvature is relatively away from the center of the vortex but as

$5x10^{-29}$ *is closer to 0 radians than 2π radians which is the center of the vortex of space −time which is doing the pulling on other space −time at the peripheries our universe is near the outer rim being close to a vacuum state over* most of the 10^10 light-years of space with galaxies occupying only a very small percentage of the space in our universe. Are photons a continuum or not?

7.3 ARE PHOTONS A CONTINUUM OR NOT?

Photons are almost massless but they do have a miniscule mass although they do show gravitational effects such as at the event horizons of black holes ; Light is bent by gravity and gravity is the curvature of space-time due to mass. Therefore light and other forms of electromagnetic radiation must have mass. This mass is the miniscule 3x10^-18 eV/c^2. Are photons composed of strings incorporating membranes such as the 1-Brane and 2-Brane which involved electric charge collection and distribution in space. These are described or expressed in conjunction with the 3-brane,4-brane and distributed throughout higher levels of branes based on the dimensionality of space. Based on String and M theory photons are comprised of strings however with Relativistic Variance(Special Relativity states that photons travel at 3x10^8m/sec in a vacuum such as the false vacuum of deep space; however this conclusion doesn't incorporate Boso- Einsteinian Condensate or the temperature correction from the Background Microwave Radiation from The Big Bang which causes all thermometers(kelvin and centigrade to express temperatures which are 2.74 degrees colder tha they actually are.). In actuality photons are trapped in a matrix of Boso-Einsteinian Condensate at tem[eratures below 2.74 degrees kelvin and are stopped only vibrating in the matrix. Based on laser studies at near absolute zero photons (Dr. Lene Hau at Harvard)are decelerated from "c" toward 0 implying that photons are all electromagnetic radiation travels at all speeds except zero. Note the value of 3×10^{-18} eV/c^2 is a non-resting mass for photons and a resting mass of 0 never occurs as photons still vibrate in the lattice of Boso-Einsteinian Condensate as a suspension. Therefore photons do have mass and are therefore composed of strings. Strings only formed after the First Event or the Time Oscillation Paradox so the statement. In the beginning there was darkness(the Infinite Parallel Planes or the Fermionic or vacuum state)and tachyons which comprised the pre-Higgs Boson and Field of the "God Particle" thus that statement appeared to be true albeit difficult to prove in th laboratory". Let there be light"incorporates electromagnetic radiation which only formed after the formation of strings after the First Event. The converse view is that photons are totally massless and immortal which means they were incorporated into space with the Infinite Parallel Planes prior to the First Event. In that case photons would be a suspension in the space of the parallel planes trapped as a matrix similar to Boso-Einsteinian Condensate ar absolute zero (0 degrees kelvin)which was the temperature of the Infinite Parallel Planes rprior to the first event;however they may have also expressed the miniscule kinetic energy between the planes to kept the D-0 branes from being in contact with each other which would have formed dimensions. So was that kinetic energy from photons or leptons and gluons? As the release of the Super-strong Force occurred with the disruption of gluons in space during the First Event(potential energy transformed into kinetic energy)it would indicate that 100 percent potential energy isn't possible except at exactly absolute zero which can never be broached with the present forms of matter and our laws of physics although matter and strings weren't formed until after the First Event.(Odd forms of matter which were recorded at below absolute zero were really breeching 2.74 degrees kelvin and not absolute zero.)So the answer as to whether photons are "immortal "and part of the space continuum is still unclear unless and until the B.M.R. can be totally eliminated from the environment of the laboratories trying to breech absolute zero.

CHAPTER 19

A SHORT ESSAY REGARDING THE EQUATION OF EVERYTHING

Basically, space-time is the circumference of the compactified form of string theory and M theory and the spiral operator operates on angular momentum with the spiral increasing from a point of infinite curvature to a flat open circumference at the upper extreme in the numerator and a decreasing infinite flat surface or manifold to an infinite curvature point in the denominator. This forms an infinite number of constants as the infinite product over an infinite number of eigen-states of energy which form $\infty \div \infty$ *which is everything except zero* and proves conclusively that nothing doesn't exist, This spiral operator is an infinite number of constants added to the sum total of all Riemann Forces of Nature x 2π.

Reynold's Number will be substituted for ∞.

$$\mathbb{R}(n)_{abcd} = \prod_{n=\mathbb{Z}}^{\text{Reynolds number}} \frac{2^{n+1} \pi \, \omega_{ij} \cdot g}{2^n \pi \, \omega_{ij \cdot i}} \cdot \frac{\Lambda_{abc} \pm \frac{1}{2} \Lambda_{qab}}{\rho_{ab}}$$

$$\rho_{ab} \equiv \hbar \rho_{ab} \equiv \pm z \, \hbar \rho_{ab}$$
to include potential energy

$$\mathbb{R}(n)_{abcd} = \prod_{n=\mathbb{Z}}^{\text{Reynolds number}} \frac{2^{n+1} \pi \, \omega_{ij \cdot g}}{2^n \pi \, \omega_{ij \cdot g}} + \frac{\Lambda_{abc} \pm \frac{1}{2} \Lambda_{qab}}{2 \hbar (\rho_{ab} + \mathbb{Z})} \cdot \frac{C}{\text{rate}}$$

$n =$ eigenstate of energy

$$\mathbb{R}(n)_{abcd} = \prod_{n=\mathbb{Z}}^{\text{Reynolds number}} \frac{2^{n+1} \pi \, \omega_{ij \cdot g}}{2^n \pi \, \omega_{ij \cdot g}} + \frac{\Lambda_{abc} \pm \frac{1}{2} \Lambda_{qab}}{2 \frac{h}{2\pi} (\rho_{ab} + \mathbb{Z})} =$$

$$\prod_{n=\mathbb{Z}}^{\text{Reynolds number}} \frac{2^{n+1} \sigma \, \omega_{ij \cdot g}}{2^n \sigma \, \omega_{ij \cdot g}} + 2\pi \left(\frac{\Lambda_{abc} \pm \frac{1}{2} \Lambda_{qab}}{2 h (\rho + \mathbb{Z})} \right)$$

$$\mathbb{R}(n)_{abcd} = \prod_{n=\mathbb{Z}}^{\pi \text{ (10 Brades } 10^{77} \text{ jonts By one}} \left(\frac{2^{n+1} \pi}{2^n} \omega_{ij \cdot g} \right) + 2\pi \left(\frac{\Lambda_{abc} \pm \frac{1}{2} \Lambda_{qab}}{\pm 2h (\rho_{ab} + \mathbb{Z})} \right)$$

Circumference ↑ eigenstate of energy

88

$$\{ 10^{77} \text{ joules} \subset \underset{\text{number}}{\text{Rayo's}} \mid \underset{\text{universe}}{\text{our}} \subset \text{multiverse} \}$$

$$C \stackrel{\text{is a}}{=} \text{subset of}$$

so

$$R^{(n)}_{abcd} = \prod_{n=\pi}^{10^{77}\text{joules}} \frac{2^{n+1}\pi\, W_{n \to j}}{2^n \pi\, W_{n \to i}} + \frac{R_{abc} \pm \frac{1}{2} R g_{ab}}{\pm 2\hbar (\rho_{ab} + \pi_{ba})}$$

← for our universe

or

$$R^{(n)}_{abcd} = \prod_{n=\pi}^{\text{Rayo's number}} \frac{2^{n+1}\pi\, W_{n \to j}}{2^n \pi\, W_{n \to i}} + \frac{2\pi (R_{abc} \pm \frac{1}{2} R g_{ab}}{\pm 2\hbar (\rho_{ab} + \pi)}$$

↑ ↑ ↑
curved Finite no. Riemann forces ∧ Nature = R
spacetime of constants
is curvature < ∞

$$\prod_{n=\pi}^{\text{Rayo's number}} \frac{2^{n+1}\pi\, W_{n \to j}}{2^n \pi\, W_{n \to i}} \subset \prod_{n=\pi}^{10^{77}\text{joules}} \frac{2^{n+1}\pi\, W_{n \to j}}{2^n \pi\, W_{n \to i}}$$

↓ ↑
Multiverse our universe

$$\{10^{77}\,C_{joules}\ \text{Rayo's number}\ |\ \text{our universe} \subset \text{multiverse})$$

$$C = \text{is a subset of}$$

so
$$R^{(n)}_{abcd} = \prod_{n=\pi}^{10^{77}\,joules} \frac{2^{n+1}\pi\,\omega_{j\to j}}{2^{n}\pi\,\omega_{j\to j} \geq 2} + \frac{R_{abc} \pm \tfrac{1}{2}\Lambda g_{ab}}{\pm i\hbar(\ell_{ab} + \Lambda_{ab})}$$

← for our universe

or
$$R^{(n)}_{abcd} = \prod_{n=\pi}^{\text{Rayo's number}} \frac{2^{n+1}\pi\,\omega_{j\to 0}}{2^{n}\pi\,\omega_{j\to i}} + \frac{2\pi(R_{abc} \pm \tfrac{1}{2}\ell g_{ab}}{\pm i\hbar(\ell_{ab} + \Lambda)}$$

↑ Riemann tensor + Noether ↑

↑ curves geometric 's circumference

↑ Finite no. of constants $< \infty$

$$\prod_{n=\pi}^{\text{Rayo's number}} \frac{2^{n+1}\pi\,\omega_{j\to j}}{2^{n}\pi\,\omega_{j\to i}} \subset \prod_{n=\pi}^{10^{77}\,joules} \frac{2^{n+1}\pi\,\omega_{j}}{2^{n}\pi\,\omega_{j}}$$

↓ Multiverse

↑ our universe

Reynold's Number will be rehabilitated to ∞.

$$\mathbb{R}(n)_{abcd} = \prod_{n=\pi}^{\text{Reynolds number}} \frac{2^{n+1}\pi\, \omega_{i\to j}}{2^n \pi\, \omega_{j\to i}} \cdot \frac{R_{abc} \pm \frac{1}{2} R_{qab}}{R_{ab}}$$

$$R_{ab} \equiv \hbar\varrho_{ab} \equiv \pm i \hbar \varrho_{ab}$$

to include helicities — quality

$$\mathbb{R}(n)_{abcd} = \prod_{n=\pi}^{\text{Reynolds number}} \frac{2^{n+1}\pi\, \omega_{i\to j}}{2^n \sigma\, \omega_{j\to i}^2} + \frac{R_{abc} \pm \frac{1}{2} R_{qab}}{i\hbar(\varrho_{ab}+\pi)}$$ $\bar{c}$ — max rate…

$n=$ eigenstate of energy

$$\mathbb{R}(n)_{abcd} = \prod_{n=\pi}^{\text{Reynold's number}} \frac{2^{n+1}\sigma\, \omega_{i\to j}}{2^n \pi\, \omega_{j\to i}} + \frac{R_{abc} \pm \frac{1}{2} R_{qab}}{\frac{i\hbar}{2\pi}(\varrho_{ab}+\pi)} =$$

$$\prod_{n=\pi}^{\text{Reynolds number}} \frac{2^{n+1}\sigma\, \omega_{i\to j}}{2^n \sigma\, \omega_{j\to i}} + 2\pi\left(\frac{R_{abc} \pm \frac{1}{2} R_{qab}}{i\hbar(\varrho+\pi)}\right.$$

10^{77} (if bounded joints by one)

$$\mathbb{R}(n)_{abcd} = \prod\left(\frac{2^{n+1}\pi\, \omega_{i\to j}}{2^n\, \omega_{j\to i}} + 2\pi\left(\frac{R_{abc}\pm \frac{1}{2}R_{qab}}{\pm i\hbar(\varrho_{ab}+}\right.\right.$$

↑ Circumference   ↑ eigenstates

CHAPTER 20

PHYSICAL COSMOLOGY; ISOTROPISM VS. AN ISOTROPISM OF OUR UNIVERSE

Based on the large scale distribution of mass and constraints on the large scale anisotropy(observational asymmetry)of the Background Microwave Radiation $\frac{\delta T}{T} \sim 10^{-5 mwit}h$ *the velocity field being* 30 h *MParsecs and the clustering of galaxies and mass*. The Sacks-Wolfe Relation (1967) has the CBR temperature inversely proportional to temp in degrees kelvin. The δ h *is measured orthogonally* to indicate quadruple anisotropy $\delta \frac{T}{T} = \delta h$ *where t* 0 $>10^4$ *such that the large scale expansion appears to be close to isotropic. The Newtonian potential* Ener gy due to gravity

is $\phi \sim \frac{G\delta m}{a} x = \frac{\frac{G}{a}x\delta m}{m} \frac{4}{3\pi} = \rho\, B(ax)^3 = \frac{1}{2\Omega H^{2(ax)^2}}$ δ x or 0.5 Ω $(H^{2(ax)2\delta x} \cdot \frac{\delta m}{m} =$

$\delta x^a a(t)$. That is $\frac{dm}{m} = \delta x^a\, (a(t))$. *With the Einstein deSitter Limit expansion scale a to* $a^{-\frac{3}{2}}$ *or H is directly proportional to a relating to the expansion to the* $\frac{3}{2}$ *power the* Gravitational potential energy *due to mass fluctuations ϕ are independent of time. This is because the density constant* indicating homogeneity of this universe grows at $\frac{\delta m}{m} = \delta x^a a(t)w$ hereby a t indicates Hubble Time and delta $\frac{m}{m}$ is the change in mass over Hubble Time. This incorporate the all important spiral fractal formula of $\int \frac{du}{u}$ where u = mass and *the limits relate to Hubble Time. The gravitational redshift* δ $\frac{v}{v} \sim \phi$ *and there are* perturbations to the CBR temperature. CBR anisotropy is caused by mass density fluctuations and like perturbations to the CBR temperature. This relates to a commoving scale at Hubble distance(x). So $\frac{\delta T}{T} \sim (H0a0x)^{2\delta x(0)}$ with a *a potential as being k. The avg length is Hubble Length or* $a1x. \frac{\delta T}{T} \sim \delta x(0)$ On an angular scale this boundary goes down to fractions of seconds of arc where one second is 1/3600 degree.2 This results in the Sachs-Wolfe Relation decreasing with a decreasing scale with respect to *δ x. Thus the CBR anisoropy shrinks on the angular scale as subdivisions reduce toward second* of arc. However over time these fluctuations may not be totally time independent and the clumpiness may change due to the anti-gravitational effect of the cosmologic constant flattening out space-time. Still as the expansive component continues to expand geometrically and the rotational component from the Big Bang reduces significantly due to Regee Slope Trajectories homogeneity appears to increase with time an anisotropism becomes more difficult to detect which is why gravity waves must be used to measure space-time curvature to locate the supermassive black hole as the location of the Big Bang and the center of mass in what appears to be an only slightly anisotropic universe. If there is no center of mass in our universe due to ISOTROPY that means that A Big Bang COULD NOT OCCUR IF IT DOESN'T ORIGINATE FROM A POINT IN SPACE. Since the syllogism that the Big Bang cannot originate from any point in space it's a false syllogism. Does this mean that there wasn't a Big Bang?The accelerating expansion of our universe can be caused by space-time pulling space-time as a current outward in all directions and the pulling space-time traveling at 2.2x10^35(c)meters/sec can be pulling our local space-time toward the center of a vortex from the outward position of the vortex where spacetime is almost totally flat(5x10^-29 radians). If dark energy is pushing space-time and space-time is pulling space-time in the same Cartesian coordinates there doesn't have to be an origin point to expansion the accelerating expansion of this universe. This gives some volition to the "Steady State Theory "which was eclipsed by "The Big Bang" however the first event or the infinite parallel plane hypothesis still holds forming a vortex with out universe being on the periphery forming from strings. despite this, this author is still agnostic about this universe being totally isotropic and homogeneous as there may be measuring defects in the red shifts as there are no clear directions where the red shifts are gravitating to. either there was a big bang and our universe has a center of mass and a rounded circular edge or there was no beginning since the formation of strings as the first event.

Our universe is near the outer periphery of the space-time continuum due to the flatness of space-time in our universe. This means that the pulling effect from spacetime in the vortex towards the center of the vortex or it's center of mass is not as strong as it will be as the center of the vortex is approached. The vortex is so huge that at the periphery a center of mass cannot be detected with our technology and with the miniscule curvature of our space-time at 5×10^{-29} radians we note the expansion due to the space-time current and dark energy plus homogeneity and isotropism as the center of mass of the system is the center of the vortex which is almost an infinite distance away from us. Based on that our clinical observations could show that our universe may have been formed from string right after the First Event without a Big Bang although Inflation is still possible. The degree of space-time curvature for an entire universe can be utilized as a partition function relative to the total space-time curvature of the space-time continuum(which is spiral asymptotic to a cone comprised of a confluence of an infinite number of 2-D-spheres or flat hyper-spherical surfaces progressive decreasing in diameter from an infinite diameter with flat space-time curvature to a point of infinite space-time curvature.)Considering a total space-time curvature of our universe as 5×10^{-29} radians as compared to an almost infinite amount of space approaching but never reaching 0 radians; space-time curvature can locate the relative position of any universe in the space-time continuum a long as the degree of space-time curvature or flatness for each segment of the space-time continuum can be determined. The center of the vortex has approaching infinite curvature and the extremes have virtually no curvature so a plot of space-time curvature versus relative position can be created. It would make sense that any "vacuum universes" with totally or almost totally flat space-time would be at or near the periphery of the space-time continuum. In the event of the existence of the multi-verse one must know the composition of each universe and mass to determine it's particular space-time curvature and therefore locate its position,

20.1 WHAT IF THERE WAS NO RELATIVE POSITION FOR A BIG BANG?

In order to have a homogeneous isotropic universe with no center of mass then any Big Bang would have either occurred everywhere or would have to keep banging throughout the expansion at every possible site in space-time. The mass distribution has a constraint of the Sachs-Wolfe Effect mentioned earlier. Mass fluctuations $\frac{\delta m}{m}, > 10^{-4}$ *Over ∞volume the mass fluctuations in the mass distribution becomes the Hubble Length*~4000 Mega parsecs as a boundary to L H or Hubble Length. The mass fluctuations are ~1 when what's termed the smoothing radius(the central path with reference to Hubble length) is reduced to 1% of Hubble Length where $dl/dt = H_0 (l)$ where l is the mean distance between conserved particles which increases with time at $H_0(l)$. H_0 is the Hubble Constant L h(Hubble Length)=c/H_0~ 4000 Megaparsecs. The recession velocity of the expansion extrapolates to "c" or 3×10^8 m/sec which is actually in space-time traveling in all directions at *πc in every and all directions while being pulled at* $2.2 \times 10^{35(c)}$ *the current of space −time heading toward the middle of the vortex formed with the Infinite Parallel Planes.*

As the Big Bang is an identifying event triggering an explosions it suggested the event must have occurred in space. If the universe has no center of mass there is no localized explosion in space and the only physical evidence that exists for this is the background microwave radiation and the WOMP which illustrated some clumpiness. There is no clear evidence that any site of the explosion had a localized center,As a consequence the is only soft evidence that such an event actually happened. This indicates that after the Time Oscillation Paradox acted on the Infinite Parallel Planes strings formed matter and anti-matter after the spin2 vector bosons occurred. This indicates that after the time oscillation acted on the infinite parallel planes that there wasn't an explosion but perhaps Inflation or a mutual repulsion between anti-particles which pushed matter and a scintilla of anti-matter outward in all discernable directions from everywhere. Of course this pre-supposed that the anti-matter may homogeneously distributed throughout the matter which caused a sequestration of the anti-matter and matter at opposite poles at a fracture point with a symmetric throat at the 0,0 point. So as a consequence either there was no Big Bang or this universe of ours has a center of mass and a discernible location in space where the Big Bang occurred. It was postulated by this author that there is a supermassive black hole not yet discovered by space-time curvature measurements which is the center of mass and that this universe is not totally isotropic and homogeneous. If this wasn't the case the formation of strings formed either the entire multiverse or just one universe which pushed out in all directions from any scintilla of anti-matter after the anti-matter was formed.

20.3 OCCAM'S RAZOR(does it help?)

The scientific precept of Occam's Razor(Ockham's Razor)states that all things being equal the simplest explanation tends to be the right one. So what is more likely,that our universe has no center of mass or gravity when it has a mass of 10^{64} kg and that a quantum bubble(containing everything) appeared out of absolutely nothing and exploded in all directions with the explosion no point of origin in space creating a universe with no edge or the universe has a supermassive black hole which hasn't been discovered yet where the rotational vector originated before the expansion and that the universe rotated around that central axis while expanding geometrically. Normally the answer should be the latter however part of Occam's Razor also states "all things being equal "which means that there aren't outside systems acting into the equation with missing data and information. An example of this is the equation 345+-----=7----. What's the answer; insufficient data. With insufficient data even in a minimum information problem it still doesn't compute with one answer. As one doesn't know if the sum is 7,70,71,700,750,7000,7546.... all the way out to 7(infinite number of digits there are an infinite number of solutions so there is insufficient data). In this equation all things aren't equal. Also the precept of an intelligent or non-intelligent observer must be present and changes reality also flunks Occam's Razor but the observer means "all things aren't equal". Sadly there may still be "insufficient data" like the addition problem to determine the correct solution. "The Omega Point "was postulated by Frank Tipler Ph.D. as that point whereby everything learnable is learned. As the storage capacity of our brains is probably under that; an individual is unlikely to reach that, however a Turing Test positive automaton with deca-byte or decillion-byte storage capacity may. As only a small percentage of the equations in nature have been discovered or derived approximately 25% scientists have a long road a head of them to fill in the "insufficient data" to answer every question in nature. It may take this to determine the full scope of any and all outside systems acting upon nature.

20.4 WAS THE BIG BANG A BIG SWIRL? The rotational component of the two sequestered poles of matter and anti-matter was maximum at Planck Time 10^{-43} seconds... If the ORIGINAL POINT WAS THE CENTER OF THE VORTEX FORMED WITH THE TIME OSCILLATION PARADOX AND IF STRINGS(including matter and anti-matter)initially swirled outward in every direction from this site with the anti-gravitational moment of anti-particle antiparticle repulsion emerging from the extreme sequestration of the spin2 vector bosons from that site after the formation of string dimensions coupled with the annihilation of matter and anti-matter negating the possible anti-gravitational effect of matter and anti-matter this combined force swirling outward at 2π *radians there* May have been sufficient force to propel our universe outward to the periphery of the vortex of spiral space-time prior to the expansion pull of space-time back toward the center of the vortex. This would be a modified spring effect but as our universe doesn't seem to reflect a bouncing effect or a recoil and since the directions would be 180 degrees from the direction back towards the center of the vortex the expansion MUST BE A RECOIL EFFECT FROM THE SPRING EFFECT OF THE BIG SWIRL OR BANG. As a consequence the location of this universe in the space-time continuum would be the MAXIMUM DISTANCE that the spring effect propelled matter and energy prior to the extreme pull of space-time back towards the center of the vortex. The only contrary evidence to this idea is that the velocity of the expansion reveals no deceleration prior to the accelerated expansion of 6.75x10^34 erg. and the accelerated expansion would be *π radians or* 180 *degrees from the inital spring effect*. This concept was in this author's first book "Mega-physics:A New Look at the Universe" and there may have been 700 googoplex of similar events with different configuration or similar configurations to that of our universe including an anti-swirl corresponding to out swirl either cohabiting the same space with a phase shift explaining overlap phenomena which can't easily be explained such as paranormal phenomena or may inhabit different space entirely. This also was discussed in this author's first book "Mega-physics,A New Look at the Universe"and this anti-swirl phenomenon would absorb any deceleration of the accelerated expansion after Planck Time by moving in the opposite direction with the same or similar magnitude. The unaccounted for mass as described by dark matter(made of baryonic particles,neutrinos,WIMPS and other "stuff" which acts as a cosmic glue) may be heterotic in nature;flipping between dimensions with the odd dimensions being directly measurable(ordinary matter) and the even dimensions only indirectly measurable(dark matter with a mass=mass of ordinary matter/i). These even dimensions may reflect with the antiswirl component of either our universe or a parallel universe coexisting in the same dimensional matrix. The 1-brane and 2-brane reflect electric charge,the electron(or positron) and it's spread or distribution as that also of any electromagnetic radiation. The 3-brane and 4-brane reflect space for ordinary matter(3 dimensional) and "flat"or 2 dimensional matter where the third dimension is string sized as in the photon. The 3- brane would reflect space for photons,etc and the 5-brane may contain dark matter with a hybrid mass which oscillates between + and − but isn't + or − making it's direct measurement very difficult except via higher gravitational measurement than projected mathematically. In the book "Megaphysics II:An Explanation of Nature" this author described dark matter as the cosmic residue from burned out anti-matter from the Big Bang like a match with phosphors being the matter and anti-matter the striking being the Big Bang and the charcoal residue on the tip of the match being dark matter. During a phase shift(see" Megaphysics:,A New Look at the Universe")it may be possible to DIRECTLY MEASURE DARK MATTER as well as any other phenomena among other dimensions including paranormal phenomena. As mentioned in the book "Megaphysics,;A New Look at the Universe"the inertial mass from the missing mass including dark matter should have slowed the accelerated expansion down, which it didn't so as a consequence the additional gravity or curvature of space-time caused by the inertial mass must be severely affected by the properties of that inertial mass. If the inertial mass doesn't slow down the acceleration it may not be in our universe or at least part of the inertial mass may not be in our universe and as a hybrid it would be the −mass oscillation that exists either in shadow dimensions or a shadow universe. The perception of our brain records images through the eye which are inverted and reduced(real images)and the brain flips them upright and adjusts the size. In the same scope, the measuring devices of inertial mass must be able to measure ALL MASS even hybrid or oscillating heterotic mass.

CHAPTER 21

HOW THE EQUATION OF EVERYTHING APPLIES TO THE BIG BANG

$R\, abcd(n) = \prod_0^\infty 2^n + \frac{1(\pi\omega\, i \to j)}{2^{n\pi\omega} j} \to i + 2\pi(R\, abc + or - \frac{1}{2R} g \frac{ab}{+} or - ih(\Lambda\, ba + \rho\, ab$.

In the case of The Big Bang as a spring effect with a countercurrent in space me from the from the center of the vortex of space —time out toward the relative location of our universe in the vortex the spiral operator reahes reaches zero when the spring was exausted before the accelerated expansion back towards the center of the vortex. This is shown as $2^{\wedge}n+1\pi\omega\, i \to j\ goes\ from \infty\ to\ 0\ and\ 2n\pi\omega\ j \to i\ goes\ from\ 0\ to\ \infty\ making\ an\ infintie\ number\ of\ constants\ except\ 0\ totalling\ up\ to\ the\ infinite\ p$product of 0. The n eigen-state of curved Lorenzian or Riemannian Space-time has flat eucidian space with gravity and anti-gravity effect divided by the energy equivalent to inertial mass or R $ab = \hbar\rho\, ab\ or\ h(\Lambda\, ba + \rho\, ab) where\ \hbar = h\frac{}{2\pi} and\ h = Planck!sConstant.\ The\ square\ root\ of\ -1\ or\ i\ relates\ to\ the\ Wave\ Function\ in\ the\ Schrodinger\ Equation\ and\ the\ = +or - i\ relates\ to\ the\ heterotic\ property\ as\ in\ the\ Wick\ Rotation.\ The\ 0\ eigenstate\ at\ the\ center\ of\ the$ vortex relates the the near infinite string dimensions after the Time Oscillation Paradox acting on the Infinite Number of Parallel Planes. In this case $2^{\wedge}n=1$ times pi times angular momentum going from the final state j to the initial state i=1/pi(omega)j.$2^{\wedge}n+1=2$ so 2(pi)omega I to j/(pi)(omega j to i=2(net total angular momentum). The net angular momentum omega I to j/omega j to i approaches zero. So in the zero dimensional state before the time oscillation paradox the angular momentum was zero but approaches infinity after the time oscillation paradox as the angular momentum goes from approaching zero to approaching infinity with energy as 10^{77} joules forming from nearly 100% potential energy into potential and kinetic energy. Planck's Constant is the energy level of electromagnetic radiation divided by the frequency of the electromagnetic radiation and therefore is the correction between inertial mass as described by r ab or the ricci tensor and the energy density of matter ρ ab $+\Lambda$ ba *which is the energy density of a vacuum. As the deBroglie Equation states* wave-length=h/mass(velocity) or h/mass(speed of light) but it states that every wave function has an associated frequency and wavelength associated with its respective mass related by Planck's Constant. The energy density of a vacuum is described by the Cosmologic Constant before the time oscillation paradox and by the energy density of matter plus the cosmologic constant after the time oscillation paradox and the energy density of matter approaches 10^{77} joules after the spring from the center of the vortex approaches the relative location of out universe in the space-time continuum. When the spring exhausted and the acceleration of spacetime as a current began approaching the center of the vortex. the expansion of the universe is occurring in constricting space-time as it approaches the center of the vortex making the expansion appear to be greater than it actually is from the standpoint of the observer who is part of the system that is expanding with increasing flattening of space-time within a vortex of constricting increasing curvature space-time so the effect is flattening or increasing curvature of space-time within and without the system in the larger system makes another current or countercurrent in space-time dampening the accelerated expansion toward the center of the vortex. As this dampening effect continues over time the expansion may continue to decrease and stop eventually ending in Heat Death(Frank Tipler Ph.D. The Anthropic Cosmologic Principle) or even a Big Crunch.

Dr. Mitchell Wick

21.1

THE REGION OF SPACE-TIME OF THE PRIMORDIAL BLACK HOLE AND THE SITE OF THE BIG BANG

R=Region of space-time R abcd=Region of space-time which is curved R abc=Region of space-time which is flat. In the case of the Big Bang the region of space-time is the center of the vortex of space-time formed from the first event. This region has infinite curvature and is constricted towards zero without reaching it. As a result R abcd in the first eigen-state is constricted to 0 with infinite curvature resulting in the equation as R abcd(1)=0 and R g ab=∞ *as R g ab is the curvature of space − time caused by mass*. Total curved space-time=total flat space-time+ or − spacetime curvature metric/mass doing the curving. In the case of the primordial black hole Rabcd(1)=∞=Rabc+∞*(infinite curvature)*/∞*(infinite mass)*=everything except zero. Curved space-time which is constricted toward zero has infinite curvature from gravity with near infinite mass in the space-time continuum. This presumes that the space-time continuum after the first event had infinite mass and infinite gravity rather than finite mass and finite gravity. As a consequence EVERYTHING EXCEPT ZERO(space-less-ness) is R abcd in the first eigen-state after the first event reveals infinitely curved space-time constricted in the case of the center of the vortex of the space-time continuum where the effect of gravity approaches infinity as does the mass which must be the total mass as infinity/infinity=everything except zero. The spiral operator at i=initial event must be the center as the inifinite product of the numerator is $2\hat{}n+1\pi\omega$ $i \to i$ *where* $n = 1$ *the eigenstate after the first event*. $2^n+1=2^2=4$. *This becomes Poisson's* Equation (∇ ω, ρ τ = $4\pi\rho$ *total* ;as the result of the numerator which is an expanding spiral operator is 4 $\pi\rho$ *where ρ is the energy density of matter. So as a consequence the angular momentum is initially the energy density of matter*. The dual vector field is ρ,ω *where ρ is the energy density of matter and ω is angular momentum. Stress Energy T ab $4\pi\rho$ 2 as ρ R g ab + ρ R ab where R g ab is gravity and R ab is inertia reveals the stress energy tensor* *times* $8\pi or 8\pi$ This is added to everything except zero and the result is clearly the entirety of the curved space-time continuum WITHOUT SPACE-LESS-NESS OR NOTHING WHICH DOESN'T EXIST. $8\pi T$ *is the solution to Einstein!sEquation of Relativistic Gravity.*

CHAPTER 22

INFORMATION EXCHANGE VIA SPOOKY ACTION AT A DISTANCE

As the solution to the Hawking Paradox is the transmission of information from a dissolving black hole event horizon like a suspension in fluid space-time like tang in water or Na Cl in water; yet Spooky Action at a Distance between photon pairs of electrons with regard to spin or charge can be over vast distances. Can this dissemination of information between electron "gasses" which spread out ad infinitum also apply to the dissemination of information from the event horizons of black holes also. If that is so the THERE WILL BE AN INFORMATION EXCHANGE BETWEEN BLACK HOLES JUST AS THERE ARE BETWEEN ELECTRONS OR PHOTONS AT GREAT DISTANCES. How can one prove that the information in one black hole contains the information in another black hole? If this is true the information from the supermassive black hole at the site of the Big Bang can be "smeared" throughout space-time and concentrated at the event horizon of other closer black holes. The answer is still gravity waves or changes in space-time curvature. Examining the ideal space-time curvature according to Schwarzchild Space-time of any black hole and comparing it with measured perturbations in space-time curvature according to the LIGO project will give a minute difference where Schwarzchild Space-time is calculated and LIGO space-time is measured. The difference is a perturbation due to information gathering from other black holes including the supermassive black hole at the Big Bang which shows more curvature due to the increased rotational moment than other black holes.

MATH CHAPTER 22.1

Space-time or R abcd(n) is $-1/2\, e^{\wedge}{-}in\, \cot\theta$ *which when the spiral operator is applied as* $2^n + 1\pi\omega\, i \to \dfrac{j}{2n\pi\omega j} \to$ *i has the information from a dissolving black hole as Hawking Radiation spuming as a quasar* Hawking Radiation relates to the Unruh Effect 3 and the equivalence principle with respect to Black Hole Event Horizons. The Schwarzchild Black Hole metric is $ds^2=(1-2M/r)\, dt^2 + 1/1-(2M/r)\, dr^2 + 2r^2 d\Omega$! *with the Unruh Effect as the Temperature* $= \dfrac{1}{2\pi\rho} = 1/$ $4\pi\sqrt{2m(r-2m)}$. *In other words As objects fall into a black hole the observer feels accelerate* In Minkowski space(flat) by the Equivalence Principle. The gravitational redshift is the square-root of the metric of space-time curvature. Utilizing the Stefan-Boltzman Constant $\sigma = \dfrac{\pi^2 k^4 \beta}{60} \hbar^3 c^2$ the Schwarzchild Radius of Black Hole event horizons is $r\, s = \dfrac{c^4}{4GM}$ Black Hole Surface Gravity is $g = \dfrac{GM}{r^2(s)} = \dfrac{c^4}{4GM}$ nd the corresponding energy is $\hbar \dfrac{g}{2\pi c} = \hbar \dfrac{c^4}{2\pi c \left(\frac{c^4}{4GM}\right)} = \hbar \dfrac{c^3}{8\pi GM}$ *for Hawking Radiation. The peak wavelength of Hawking Radiation is 16 times the Schwarzchild Raduis.*

APPENDIX

BRANES;a string is a one-brane which couples to a background second degree tensor. zero-branes are ten dimensional building blocks for space in the pre-big bang epoch. the second degree tensor is purported of negligible mass as indicated by the zero-brane. the source of the background second degree tensor is r uv where the Integral of d to the d power of x where x is the string or onebrane In d dimensions applies to r u v g u v where g u v is the Metric acting on r u v for the zero-brane with respect to x which Is the one-brane. R u v is the second degree tensor upon which The metric g u v acts. In four dimensions a monopole is dual to Two electrons acting on a zero-brane. In 10 dimensions a string Is analogous to a five-brane based on p-brane potentials. This Involves dual fields such as a tensor r=r* from ra1...a N=r8b1...b n.p-branes are encircled by a hypersphere which Relates to m theory being compactified (curled up)by a circle for Typeiia strings.

the charge of a p-brane is based on $Q = \int * R \; from \; limit \; S \; d - p - 2$ *for electric charges and* $Q \int_{sp+2} R$ *for electromagnetism. P − branes tie in with the potential involved with permutations of a field tensor. p brane tensors ar*E associated. With a tensor of the pth rank r a1...a p and electric and magnetic charges can be associated with p-branes with superalgebra. Dzero-branes represent the vacuum state. although indicated as ten dimensional building blocks of space they actually are zero-dimensional. One-branes represent strings which are two dimensional or possibly one dimensional. If all the dimensions in a system or universe are conserved such that the total number of dimensions are constant; then zero branes would have to be ten dimensional in the vacuum state. Six dimensions for CalabiYau

Manifolds and four dimensions of space-time. As the "c" boundary is approached infinite mass with reducing length and width occur when length becoming infinite. In this case width and height approach zero but do not reach it and become infinitely small curled up and compactifed. In general although showing duality between different systems which are abelian membranes are described by the forces involved with mass or energy associated with the membrane with reference of n-dimensional space where n dimensions would have n-1 membranes or n=1 brane.

FLAT OR MINTKOWSKI space is described mathematically as the line element or $ds2=dx2+dy2+dz2-c2dt2$. flat space-time is space-time without any curvature in other words a vacuum state here r g a b=0 which indicates that the space-time curvature metric=0 and therefore gravity=0 in the vacuum state.

CURVED SPACE-TIME IS GENERALLY DESCRIBED BY $ds2=e-k|r|(dx2+dy2+dz2-c2dt2)+dr2$ where r=space-time curvature metric described by tensor as R g a b. R g a b or r is determined by the inertial mass of the object doing the curving and the curving is performed by bosons and possibly gravitons or fermions. Spiral space-time has k=-i(the square root of-1) to the n cotangent theta power as suggested by Dr. Roger Penrose and proposed by this author.

MANIFOLDS

The simplest manifolds are cartesian spaces where a manifold structure or surface in terms of topologies is r to the d power with what's called an identity map Rd implies R d. The coordinate functions of this map are cartesian coordinates. If coordinates are a I; R d is the manifold of the standard Cartesian coordinates. a i=ax+ay+az and R to the d power is the tolological expression of the standard manifold or the Cartesian Coordinate system. If a manifold is imbedded in another manifold it is a submanifold. On a string basis submanifolds can be orbifolds or Calabi Yau manifolds which are submanifolds for spiral manifolds for asymptotically flat but curved space-time on a macrostatic surface which is expanding and simultaneous rotating as at black hole event horizon.

RIEMANNIAN CURVATURE

A riemannian space is the space coordinized by xi(power) with a fundamental form of the Riemannian Metric g l jdx l dx j where g=(g ij) obeys the metric tensor.g is of differentiability class C2(all second order partial derivatives of g l j exist and are continuous. g is symmetric g l j=g ji;g is nonsingular |g l j| doesn't equal 0. The differential form and distance from g isn't variant with regard to changes in coordinates.

R I j k I=g I ir(Rr superscript with jkl as a subscript where R jkl with ias a superscript is the Riemann tensor of the second kind. The Riemann Tensor of the first kind is R l j k l=$\dfrac{\Gamma jki}{xk} - \dfrac{\partial \Gamma jki}{xi} + \Gamma ilr \Gamma jk$ with r as a superscript + $\Gamma ilr \Gamma jk$ with r as a superscript − $\Gamma ikr \Gamma ji$ with r as a superscript. Here Γ is a Christoffel symbol or the derivatire of a tensor. Above is $Rijkl - \dfrac{\partial \Gamma jli}{\partial xk} - \dfrac{\partial \Gamma jkl}{dxl} + \Gamma ilr \Gamma jk$ with r as superscript − $\Gamma ikr \Gamma jl$ with r as superscript. Skew Symmetrys involve Bianchi's Identity R ijkl + Riklj + Riljk = 0 skew symmetry is R i j k l = −Ri j kl and second skew symmetry is R ijkl = −R i j l k with R j k l with i as superscript = −Rj l k withi as a superscript. Block symmetry is R i j k l = R k l i j. These symmetry properties must fir with the n2(n2 − $\dfrac{10}{12}$ COMPONENTS OF THE RIEMANN TENSOR(R i j k l) where the diagnal tensor without s. PTR

Rijkl=g l iRjkl is subscript and l as superscript in the diagonal metric te tensor calculations for the Riemann Metric gives six cases R one R 212 and 1 R 313 and 1 R 323 and 1 R 213 and 1 R 232 and 1 R 123 and 1 which proves with the partial derivatives of the Christoffel symbols of tensors according to the previous formulas give R l j k l=0 for all l j k and l indicating the summation of all Riemann forces and space is zero. The math of all these combinations is very difficult to reproduce by typing.

GLOSSARY

Abelian: equations having a coefficient or variety in a specific group,g,,algebraic number fields,tensors of the same degree or cohominy group

Anisotropic: not isotropic,lacking observational symmetyry

Anti-symmetric: tensors or vectors that are equal but opposite and can therefore partially cancel or cancel

Aymptotic: that which approaches a level or degree but never reaches it;asymptotic flatness appears without curvature but doesn't reach it

Bianchi's Identity: The identity of groups of Riemannian 4 space that is antisymmetric and Abelian and cancels each other out of being equal but opposite

"The Big Bang"A theory proposed describing a Friendman type I open expanding f;at universe with is homogeneous and isotropic

"The Big Swirl " A Big Bang with a progressively decreasing rotational vector from an infinite curvature point of space-time to asymptotic flatness

Black Hole: collapsed matter from a neutron star or galaxy with extreme curvature of space-time at the central nexus due to extreme gravity of of the spiral space-time

Calabi Yau Manifold: a surface which represents a relative isotropic portion of spacetime with a puckering to accommodate multiple dimensions considered a twisted variant of the orbifold

Choas: absolute disorder

Chiral: a mirror image or absolute symmetry

Closed string: a two or one dimensional building block of matter from energywith movements in 10 or 26 dimensions without breaking the string

Compactified: when every point of the dimensions are curled up mathematically making the size approach zero. First determined by Kaluza and Klein

Conformal Space:when every point in space relative to every other point maintains its relative position regardless of what the space is doing

Dark Matter; an indirectly measured mass causing perturbations in gravity(the curvature of space-time)caused by mass. Acts as cosmic glue containing possibly baryonic particles and neutrinos

Event Horizon: area where a black hole is perceived by easurements
Entropy:degree of disorder

Entropy: degree of disorder

Ex nihilo: out of nothing

M(Membrane)theory: the 5 dual string theories into one massive theory of everything which incorporates membranes which vibrate and incorporate all energy and matter

Isoropic: observational symmetry

Geodesic: a unit of space-time

Gravity: the curvature of space-time caused by mass;actually an effect not a force

Membranes: a description of matter in terms of energy states with stress energy densities described in the number of states with regard to dimensions N; number of dimensions in N dimensional space

Open string: a two or one dimensional bulding block of matter with movements in a multidimensional plane

Orbifold: space-time manifold in an open twisted cone configuration utilized in string theory

Relativity: the behavior of matter and energy with regard to other matter and energy;energy and space have a different vantage point from other matter and energy including stress energy,time and mass with changes regarding relative velocity

RicciTensor: that tensor which represents inertial mass or resistance against pull or push

Riemann Forces: all strong and weak forces in nature

Riemannian Space: Mintowski space with Riemann curvature of space-time caused by mass. Flat space if no mass is present

Scalar: the magnitude compone t of a vector or tensor with regard to direction

Six dimensional string manifold: curled up closed strings in configuration according to Kaluza and Klein which is 10-33 cm and may be Calabi Yau Manifolds

BIBLIOGRAPHY

1. Peat,F. David. Superstrings and the Search for the Theory of Everything. Yang Mills Forces p.114
2. Kaku, Michio. Strings, Conformal Fields and M theory. Ising Model p.176-78
3. Wald, Robert. General Relativity. Chicago,Ill. University of Chicago Press. 1984 4. CPT THEOREM; Quantum Field Theory Kaku, Michio
4. Metric tensor(General Relativity)Wikipediaa and Spacetime. en.m.wikipedia.org.spiral Space-time Einstein 1912 Fractal Time. p.108-109 Braden, Gregg 2009 Library of Congress HAWKING RADIATION. Wikipedia
5. Peat,F. David. Superstirngs and the Search for the Theory of Everything. p.106-107. Calabi Yau Manifolds
6. Kaku, Michio. Quantum Field Theory. Renormalization Actions in Quantum Field Theory
7. Peat,F. David. Superstrings and the Search for the Theory of Everything.
8. Kay, David C. Tensor Calculus p.129 Osculating Plane
9. Kaku, Michio. Strings,Conformal Fields and M theory.
10. Peebles, P.J.E. Principles of Physical Cosmology.p
11. Chang, Alan. HAMILTON JACOBI EQUATIONS UNIVERSITY OF CHICAGO 2013. Zeno's paradox: The Math Forum at Drexel University
12. Tipler, Frank j.The Physics of Immortality
13. Godel, Kurt. Godel's Incompleteness Theorems en.m.wikipedia.org
14. Randall, Lisa. Warped Passages
15. Green, Brian The Elegant Universe. and
16. Wick, Mitchell Albert. Megaphysics, A New Look at the Universe.
17. Kay, David. CTensor Calculus.
18. Kaku, Michio. Strings, Conformal Fields and M Theory.
19. Wikipedia.Electronen.wikipedia.org/wiki/Electron
20. Spooky Action at a Distance Quantum Entanglement Wikipedia. Or en.wikipedia.org/wiki/Quantum entanglement

21. Hau, Len. Harvard Research circa 2003.

BIBLIOGRAPHY

Barrero, John D. The Anthropic Cosmological Principle. Oxford England. Oxford Press. 1986
Brade, Gregg. fractal Time 2009 Library of Congress.
Greene, Brian. The Elegant Universe. NewYork. Vintage Books editor Random Press.1999
Hawking, Steven and Penrose, Roger. The Nature of Space and TimePrinceton, N.J; Princeton Science Library 1996
Kaku, Michio. Quantum Field Theory. A Modern Introduction. Oxford university Press. 1993
Kaku, Michio. Strings, Conformal Fiels, and M theory 2nd edition. Springer Press. 2000.
Kay, David C. Tensor Calulus Schaum's Outline Series. N.Y. McGraw Hill 1998.
Peat, F. David. Superstrings and the Search for the Theory of Everything. Chicago. Contemporary Books 1998
Peebles, P.J.E. Principles of Physical Cosmology. Princeton Series in Physics. Princeton University Press 1993
Wald, Robert m. General Relativity. Chicago, Illinose. University of Chicago Press 1984
Wikipedia: on lin encyclopedia.
Randall,. Lisa. Warped Passages HarperCollins Publishers. N.Y.2005
Tipler, Frank J. Physics and Immortality. Anchor Books division of Random House.1993

www.ingramcontent.com/pod-product-compliance
Lightning Source LLC
Chambersburg PA
CBHW030817180526
45163CB00003B/1329